Invaluable Invertebrates and Species with Spines

Inspire the next generation of zoologists with this award winning 30-lesson interdisciplinary science unit geared toward second and third grade high-ability students.

Using problem-based learning scenarios, this book helps students develop the vocabulary, skills, and practices of zoologists as they conduct research and solve real world problems. Students will gain an in-depth understanding of how the animal kingdom is structured, create an innovative zoo exhibit containing an entire ecosystem for a vertebrate animal of their choosing, design invertebrate animal trading cards, and much, much more. Featuring detailed teacher instructions and reproducible handouts, this unit makes it easy for teachers to adjust the rigor of learning tasks based on students' interests and needs.

Aligned with Common Core State Standards for English Language Arts and Mathematics plus the Next Generation Science Standards, gifted and non-gifted teachers alike will find this expedition into the animal kingdom engaging, effective, and highly adaptable.

Jason S. McIntosh, Ph.D., is an experienced educator of 25 years and a passionate advocate for gifted education.

GRADES 2–3

Invaluable
Invertebrates and
Species with Spines

Inquiry-Based Science Lessons for Advanced and Gifted Students in Grades 2–3

Jason S. McIntosh

National Association for Gifted Children **CURRICULUM AWARD WINNER**

Routledge
Taylor & Francis Group

NEW YORK AND LONDON

Cover image: Getty Images

First published 2023
by Routledge
605 Third Avenue, New York, NY 10158

and by Routledge
4 Park Square, Milton Park, Abingdon, Oxon, OX14 4RN

Routledge is an imprint of the Taylor & Francis Group, an informa business

Library of Congress Cataloging-in-Publication Data
A catalog record for this title has been requested

ISBN: 978-1-032-36977-8 (hbk)
ISBN: 978-1-032-36974-7 (pbk)
ISBN: 978-1-003-33474-3 (ebk)

DOI: 10.4324/9781003334743

Typeset in Chapparal Pro
by Deanta Global Publishing Services, Chennai, India

Access the Support Material: www.routledge.com/9781032369747

Table of Contents

Detailed Description of the Unit

Rationale

Kids are naturally curious about animals and enjoy watching, reading, writing, and learning about them. This unit is different from the typical second or third grade animal unit in that it incorporates advanced content, offers multiple problem-based learning scenarios for students to engage with, and teaches students to adapt the vocabulary, attitudes, and practices of real zoologists. In addition, this unit is interdisciplinary and uses the twin concepts of *structure* and *function* to tie it all together.

How to Differentiate Using this Unit

This unit provides accelerated instruction and advanced content. The variety of resources, learning activities, and products included provide options for adjusting the rigor based on the needs and interests of the students. A pretest has been included to aid in determining the degree of prior knowledge students already possess regarding

zoology, the problem-based learning method, and the twin concepts of *structure* and *function*. Daily reflection activities have been included at the end of each lesson in the form of journal prompts to help the teacher identify any misconceptions students might have and aid in planning for the next day's lesson. If at any time a student finishes an activity early, the choice menu included in Lesson 28 can be pulled and used as an anchor or extension activity. Every hands-on activity, research project, and group discussion provides additional opportunities to assess how students are progressing and to what extent the unit is meeting their needs. Lastly, the use of flexible grouping for instruction, as well as strategic construction of the problem-based learning workgroups throughout, are recommended.

Goals and Outcomes

▶ Concept goals: To better understand the world around them. The students will be able to:
  ▷ Understand and use various grouping structures to classify and organize things into like groups.
  ▷ Appreciate that grouping structures change over time when the items being grouped are better understood.
  ▷ Recognize the connections between physical structures and their individual functions.

▶ STEM goals: To form an accurate understanding of the animal kingdom while developing the skills and attitudes of a zoologist. The students will be able to:
  ▷ Differentiate between habitats and ecosystems.
  ▷ Compare and contrast various vertebrate and invertebrate phyla.
  ▷ Identify unknown animals using dichotomous keys.
  ▷ Create an accurate model of a particular animal's ecosystem while explaining the interactions that take place within it.

▶ Humanities goals: To value the contributions of scientists (e.g., zoologists, biologists, environmentalists, etc.) from the past and develop new innovative ideas for sharing the planet with our animal neighbors. The students will be able to:
  ▷ Research a scientist from the past and describe the focus of their research.
  ▷ Brainstorm creative ways they can reduce the impact of humans on the planet.

▶ Process goals: To develop critical thinking and problem-solving skills. The students will be able to:
  ▷ Use a Need to Know Board while conducting research.
  ▷ Work collaboratively to analyze information and generate solutions to novel problems.
  ▷ Reflect on their own learning and embrace constructive feedback.

Connections to Standards

This unit aligns to the Common Core State Standards (CCSS) for English Language Arts and Mathematics, as well as the Next Generation Science Standards (NGSS). To see specific standards addressed, the end of the unit includes a CCSS alignment chart and an NGSS alignment chart.

This unit also includes connections to the National Association for Gifted Children's (2019) Pre-K–Grade 12 Gifted Programming Standards, including the following:

- ▶ 1.6. Students with gifts and talents benefit from meaningful and challenging learning activities addressing their unique characteristics and needs.
- ▶ 2.3. Students with gifts and talents demonstrate advanced and complex learning as a result of using multiple, appropriate, and ongoing assessments.
- ▶ 3.1.1. Educators use local, state, and national standards to align and expand curriculum and instructional plans.
- ▶ 3.3. Students with gifts and talents develop their abilities in their domain of talent and/or area of interest.
- ▶ 3.4. Students with gifts and talents become independent investigators.
- ▶ 3.4.1. Educators use critical thinking strategies to meet the needs of students with gifts and talents.
- ▶ 3.4.2. Educators use creative thinking strategies to meet the needs of students with gifts and talents.
- ▶ 3.4.3. Educators use problem-solving model strategies to meet the needs of students with gifts and talents.
- ▶ 3.4.4. Educators use inquiry models to meet the needs of students with gifts and talents.
- ▶ 4.1.1. Educators maintain high expectations for all students with gifts and talents as evidenced in meaningful and challenging activities.

Examining Zoology

Objectives

▶ Students will complete a pretest.
▶ Students will discuss the differences between biologists and zoologists.

Materials

▶ Pretest
▶ Blank paper and pencil or a notebook for each student

 DOI: 10.4324/9781003334743-1

Assessments

▶ Pretest
▶ Performance task
▶ Journal prompt

Procedures

1. Greet the students and introduce the title of this unit. Ask the students to predict what they are going to be learning about.
2. Divide the students into small groups and ask them to introduce themselves to each other by listing their favorite animal.
3. Challenge each group to list and record on paper as many animals as they can in the next three minutes.
4. At the conclusion of the activity, ask the students to count how many are on their list and share this with the class. Each group should keep this list in a special place until the end of the lesson.
5. Tell students that a **biologist** is a person who studies living things. Ask each student to name one thing that is alive.
6. Explain that not all living things are actually animals, however (e.g., plants, fungi, bacteria, etc.). Poll the students to see if anyone knows what you call a person who studies animals (i.e., **zoologist**).
7. Tell students that you would like to find out what they already know about zoology. Distribute the pretest while emphasizing it will not be counted for a grade. Explain that the test will only be used to help you teach them, and to show growth at the end of the 30 lessons.
8. Collect the pretests for scoring upon completion.
9. If time permits after the test, explain that one important task zoologists have is to classify (or sort) animals into like groups.
10. Ask students to pull out their group's list of animals and to sort them into groups according to any trait they wish.
11. Provide time for each group to share how they sorted their animals.
12. Challenge students to regroup the same animals in a completely different way.
13. Collect the original student-generated list of animals to be used in Lesson 3.
14. End the class period by asking the students to complete the following journal prompt: "If I could have any animal as a pet, it would be_____ because_____".

TEACHER'S NOTE

Before the next class, assess the students' pretests and journal prompts to determine the degree to which students are familiar with the concepts of structure and function, as well as other important concepts related to zoology.

Invaluable Invertebrates and Species with Spines

Pre/Post-Test

Name _____   Date _____

Directions: This test is not for a grade. Please answer the questions you can to the best of your ability and skip those you cannot.

1. What does the word **structure** mean?

   _____

   _____

2. What makes something an **animal?**

   _____

   _____

3. Label each animal as an amphibian, bird, fish, mammal, or reptile using the key shown here (**A** = amphibian / **B** = bird / **F** = fish / **M** = mammal / **R** = reptile).
   a. Eagle _____
   b. Shark _____
   c. Gorilla _____
   d. Frog _____
   e. Whale _____
   f. Snake _____
   g. Giraffe _____

4. Write **five** examples of **invertebrates** in the box below:

   _____
   | |
   | |
   | |
   | |
   |_____|

5. What is the difference between a **habitat** and an **ecosystem**?

   _____

   _____

6. How are these three animals **alike** and **different** from each other?

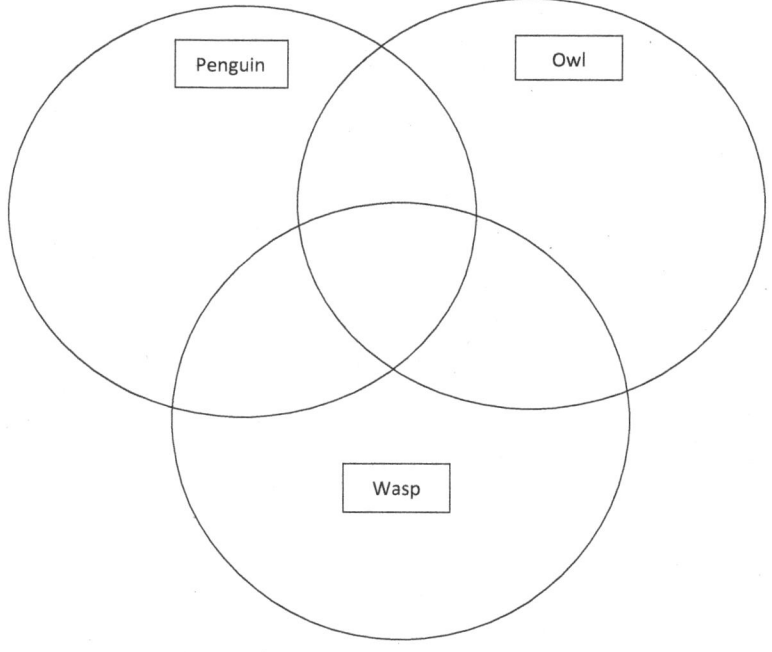

7. What do you call a **scientist** who studies animals? _____

8. Name **one** animal and describe how it has **adapted** to its environment?

    _____

    _____

    _____

9. When would you use a **dichotomous key**?

    _____

    _____

    _____

10. Name **two** parts of an **animal cell** below:

    ▶

    ▶

11. Circle the **one** answer below that is an example of an **instinct** and NOT a learned behavior:
     a. I touched the worm and it curled into a ball.
     b. The baby ducklings jumped into the pond after their mother.
     c. When the bell rings, the dogs come running for their dinner.

12. Match the terms with their definitions by writing the letter of the answer on the line.

     _____ Estivate   **A.** An animal that gets eaten.
     _____ Communal   **B.** An animal that uses energy to control body temperature.
     _____ Hibernate   **C.** An animal that eats everything.
     _____ Solitary   **D.** An animal that sleeps when the weather gets hot and dry.
     _____ Carnivore   **E.** An animal that lives in groups.
     _____ Omnivore   **F.** An animal that spends the winter sleeping.
     _____ Prey   **G.** An animal that eats meat.
     _____ Endotherm   **H.** An animal that lives alone.

13. Write **one** question you have about animals below.

LESSON 1

6

Why So Weird?

Objectives

▶ Students will create a class definition for the universal concepts of *structure* and *function*.

▶ Students will predict and then confirm the purpose of various animal adaptations.

Materials

▶ *Structure and Function* fishbone handout

▶ *Why Do They Look So Weird?* slide deck: https://docs.google.com/presentation/d/1D9mPL4Q78186xcXkkPnF_Uzybx5_mitLfBRVYk_WYkk/edit?usp=sharing

▶ *Why Do They Look So Weird?* data collection tool

DOI: 10.4324/9781003334743-2

Assessments

▶ *Structure and Function* fishbone handout
▶ *Why Do They Look So Weird?* data collection tool
▶ Journal prompt

Procedures

1. Review major concepts from last class period (e.g., zoology versus biology, etc.).

2. Explain to students that we will be connecting everything we learn in this unit to the big ideas of *structure* and *function*.

3. Ask students to recall how they defined the word *structure* on yesterday's pre-test and share it with a partner. Do not confirm or deny the correctness of their replies.

4. Give students a copy of the *Structure and Function* fishbone handout. Ask them to write words, phrases, and examples that come to mind when they hear the word structure. Each new item should be written next to a different prong on the fishbone skeleton.

5. Ask each group to share one or two things they wrote down.

6. Work together as a group to craft a class definition. One example might be *how something is arranged or grouped*.

7. Write this class definition on an anchor chart and display it prominently in the room.

8. Explain to students that they will be looking at three different types of structures in this unit:
   a. grouping structures for animals (e.g., vertebrate, invertebrate)
   b. internal structures of animals (e.g., parts of an animal cell)
   c. external structures of animals (e.g., body adaptations, habitats)

9. Repeat the process above with the word *function*. A sample class definition might be *the purpose of something or how it is used*.

10. Once a class definition has been crafted and displayed, ask students to apply this definition while thinking like a zoologist. Open the Google Slide presentation above and distribute the *Why Do They Look So Weird?* data collection tool.

11. Challenge students to choose three of the animals included in the presentation and predict why they have these structures and how they are used. Students' predictions should be recorded before moving on to the next step.

12. Provide time for students to research the truth online or using reference materials. These final conclusions are to be recorded, as well as the sources of the information, on the data collection tool.

13. Collect the handout and ask students to complete the journal prompt below to end today's lesson: "Imagine you could add a third hand, wings, or other structure to your body. What would it be and how would you use it?".

TEACHER'S NOTE

Your students may or may not have experience using online search engines or the index in a print encyclopedia or reference book. You may need to provide a short tutorial and best practices for completing these tasks.

Why Do They Look So Weird?

Name _____   Date _____

Directions: Open the Google Slide deck called *Why Do They Look So Weird?* Choose three of the animals and make an educated guess as to why each has developed the unusual feature or behavior shown. Record your hypothesis and then conduct research to uncover the truth. Record your findings and where you found the information in the chart below.

Animal's Name	Hypothesis	The Truth	Sources

Structure and Function

Name _____   Date _____

Directions: Think about the words *structure* and *function*. Use the fishbone diagrams to list what you know about these concepts and then write your own definitions in the space provided.

What is STRUCTURE?	*What is FUNCTION?*
Define the word STRUCTURE below:	**Define the word FUNCTION below:**

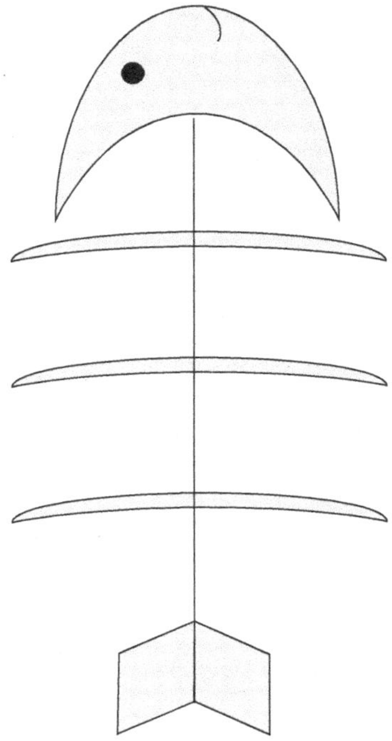

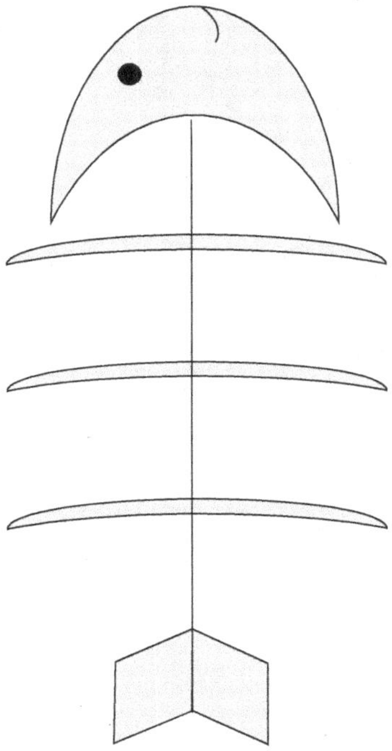

LESSON 2

How to Recognize an Animal When You See One

Objectives

▶ Students will widen their understanding of what makes something an animal.

▶ Students will discover the five kingdoms to which all living things belong.

Materials

▶ *Five Kingdoms* chart

▶ Funnel or picture of one

▶ Animal Cell diagram (or magnet set from Learning Resources): www.learningresources.com/giant-magnetic-animal-cell

▶ *The Invertebrates' Plea!* poem by Jason S. McIntosh

 DOI: 10.4324/9781003334743-3

Assessments

▶ Exit ticket

Procedures

1. Review major concepts from last class period (e.g., definitions of structure and function, etc.).

2. Ask students to imagine a visitor from another planet just walked into our classroom. The alien creature has a lot of questions about Earth, but the biggest is "What is an animal?". Give students several minutes to talk with a partner and decide how they would answer this important question.

3. Give students a chance to share their answers with the larger group. Do not confirm or deny the accuracy of their answers.

4. Pass back to students the list of animals they generated on day one of the unit. Ask students if any group of animals might be missing. If so, allow students to add to their lists using a different color of ink or marker. Set this list to the side until later in the lesson.

5. Now, ask students to put on their zoologist hats again and think like a scientist. Display the *Five Kingdoms* chart while covering up everything except for row one.

6. Show students a funnel or picture of a funnel and ask what it is for. When students have shared their thoughts, explain that a funnel is broad at the top and narrow at the bottom.

7. Explain that the broad top of a funnel is like the top row of this chart. All living things can be grouped into one of these five big groups or, as Carl Linnaeus called them when he created this structure of grouping animals in 1735, *kingdoms*.

8. Next, uncover the second row to give a few examples of what types of living things belong in each kingdom.

9. Explain that we will now look at what truly makes something an animal versus a plant, monera, protist, or fungi.

10. Before going on, make sure students know that the smallest unit of any living thing (all five kingdoms) is called a cell. Show students a diagram of an animal cell and explain the function of each internal structure. See the "Teacher's Note" for this lesson for a way to explain this easily to young students.

11. Display row three of the chart next. Point at the nucleus on the animal cell diagram and explain that this chart tells us that the living things in four of the five kingdoms have a defined nucleus like the animal cell. Ask which kingdom does not.

12. Display row four of the chart and explain the difference between a cell wall and a cell membrane (i.e., cell walls are rigid and do not allow things to pass through them. Cell membranes are permeable). Ask students which one group has a cell membrane and not a cell wall.

13. Display row five of the chart. Ask students what the prefixes *multi* and *uni* mean. If they do not know, mention words like multiplication or unicycle as a hint. Explain that *multi* means *many* and *uni* means *one*. Ask students which living things are made of more than one cell.

14. Display row six of the chart. Explain that every living thing needs some form of food to survive. Introduce the words *autotrophic* (i.e., make their own food) and *heterotrophic* (i.e., need to find an external source of food). Translate this by saying, "What this tells us is that monera, protists, and plants do not need to hunt or go to the grocery store. They can make their own food inside of their own bodies".

15. Lastly, display the final row. Ask students what the word *mobile* means (i.e., can move on its own). Mention words like *automobile* or *snowmobile* as a hint. Ask students which kingdoms have creatures that can move when they need or want to move.

16. Ask students to look at the list of animals they generated on day one once more. Tell them to remove anything they might have listed that could be found in columns one through four. Since we are studying animals, we will only be concerned with column five.

17. Next, draw students' attention to the word *invertebrate* in the *examples* row for the animal kingdom. Poll students to see if anyone knows what that means.

18. Read the poem called *The Invertebrates' Plea!* to further explain what an invertebrate is.

19. Tell students we will delve deeper into these groups in the next lesson.

20. As an exit ticket, ask students to revisit the description they provided to our alien visitor and create a more accurate definition of an animal using the chart we discussed today.

TEACHER'S NOTE

One way to explain the parts of an animal cell to students is to use the metaphor of a city. See the list below for the corresponding features:

- Cell membrane – boundaries of the city
- Nucleus – the mayor of the city
- Cytoplasm – the available space within the city
- Mitochondria – the energy plants for the city
- Centrioles – the concrete used to make the foundations for the city's buildings
- Lysosome – the waste treatment plant for the city
- Vacuole – the storage containers for the city
- Endoplasmic reticulum – the highways and roads for the city
- Ribosomes – factories for the city

Five Kingdoms

Examples	Monera	Protists	Fungi	Plants	Animals
	Bacteria	*Algaes, amoebas, ciliates*	*Yeasts, molds, mushrooms*	*Trees, herbs, ferns*	*Vertebrates, invertebrates*
Nucleus	No defined nucleus	Nucleus			
Cell Walls	Yes				No
Multi or Unicellular	Unicellular	Mostly unicellular	Multicellular or unicellular	Multicellular	Multicellular
Origins of Food	Autotrophic	Autotrophic and heterotrophic	Heterotrophic and saprotrophic	Autotrophic	Heterotrophic
Movement	Mobile	Can be mobile	Non-mobile	Non-mobile	Mobile

The Invertebrates' Plea!

By Jason S. McIntosh, Ph.D.

Some people don't think we're important.
That's something I'd like to dispute.
We may not have spines,
But we're doing just fine.
Even though we can be quite minute.

We outnumber mammals and reptiles.
Birds, fish, plus amphibians too.
We are quite diverse.
And we'll fit in your purse.
But we're scared of the sole of your shoe.

We're grouped in six different classes.
Like mollusks and sponges and worms.
But before we are through,
Give the insects their due,
And all starfish and jellies that squirm.

We hope that you will be our ally.
You'll speak up when others will say,
"Worms, bugs, and shellfish …
don't matter." They wish!
For our roles you must not underplay.

Next time that you walk down the sidewalk.
Or run cause you're late and it's a schlep.
Remember we're small.
And that most of us crawl.
So, make sure that you watch where you step!

People Puzzles

Objectives

▶ Students will use a dichotomous key to identify unknown avatars.

▶ Students will be introduced to the five main vertebrate animal groups.

Materials

▶ *People Puzzle Dichotomous Key* handout

▶ Avatars 1–4

▶ *The Big Five* poem by Jason S. McIntosh

DOI: 10.4324/9781003334743-4

Assessments

▶ *People Puzzle Dichotomous Key* handout

▶ Journal prompt

Procedures

1. Review major concepts from last class period (e.g., vertebrate versus invertebrate, parts of an animal cell, etc.).

2. Ask students to divide a piece of paper in half and label column one *vertebrates* and column two *invertebrates*. Give students two minutes to list as many vertebrates as they can. Next, give students two minutes to list as many invertebrates as they can.

3. Conduct a think/pair/share by asking students to compare their lists with a classmate.

4. Poll the students to see if anyone's partner listed an animal they had never heard of.

5. Pose the following question: "What might zoologists do if they come across an animal they do not recognize?". Give students time to brainstorm ideas.

6. Explain to students that zoologists often use what is called a dichotomous key.

7. Write the word *dichotomous* on the board. Explain that this word comes from a Greek word that means "a cutting in half".

8. Next, write the word *key* next to it. Ask the students to list all the types of keys they can think of (e.g., map key, car key, house key, answer key, etc.).

9. Distribute the *People Puzzle Dichotomous Key* handout listed in the materials section.

10. Point out to students that there are two branches each time a choice needs to be made.

11. Give students copies of the four avatar cards.

12. Model how to use the dichotomous key 1 with avatar 1 for the class. Next, let students use the dichotomous keys 1 and 2 to uncover the identities of the other three avatars, as well as the winner of the contest. Discuss the correct answers and the process together.

13. Explain to students that we will use a dichotomous key to identify various vertebrate animals in the next class period.

14. Read the poem *The Big Five* to students and ask them to pay attention to the characteristics of each vertebrate animal group.

15. Journal prompt: "Which of the five vertebrate animal groups do you know the least about? What would you like to know about them?".

TEACHER'S NOTE

An optional extension activity might be to have students create their own avatars. Students could then create a dichotomous key to sort them into groups. If students are not ready for this level of challenge, the teacher could model how to sort students into groups as a whole class demonstration.

People Puzzle Dichotomous Key

Name _____   Date _____

Directions: Use this dichotomous key and the four avatar cards to discover the identities of the four unknown people. Then, figure out which person won first place in a contest using the second dichotomous key.

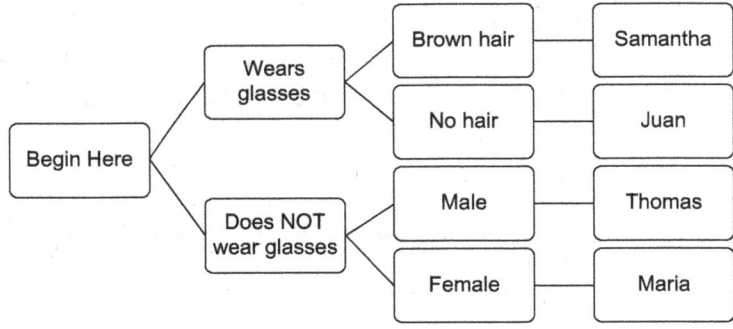

Avatar 1 = _____

Avatar 2 = _____

Avatar 3 = _____

Avatar 4 = _____

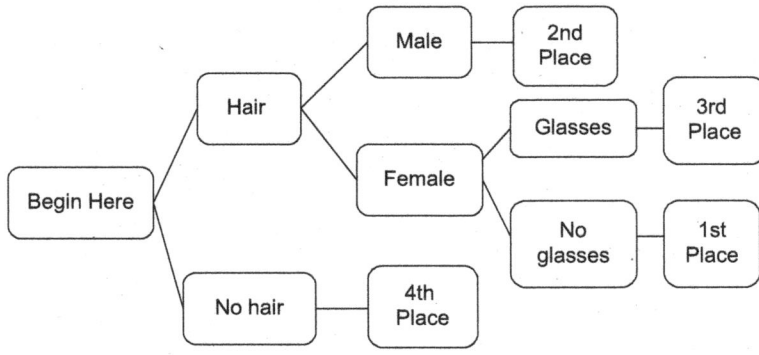

Who came in first place? _____

Avatars

Avatar 1

Avatar 2

Avatar 3

Avatar 4

LESSON 4

The Big Five

By Jason S. McIntosh, Ph.D.

When you hear "animals",
I'm sure your mind does go,
To the big five groups,
Shown in the list below.

MAMMALS are hairy,
Give live birth to their kin.
Mammals have warm blood,
A source of milk built in.

BIRDS have many feathers,
And a pair of wings.
Birds are warm blooded,
Hatch a chick that sings.

REPTILES are scaly,
And have really dry skin.
Reptiles have cold blood,
And from an egg begin.

FISH are also scaly,
And breathe air with gills.
Fish lay eggs in water,
Have cold blood that chills.

Last but not least now,
AMPHIBIANS we'll greet.
They have moist skin and
Cold blood and webbed feet.

But, I'd say if asked,
"What's most interesting?"
Amphibians can live on
The land or in a spring.

Those are the big five groups
Of animals with spines.
I hope you liked this poem,
But, it's time to say goodbye!

LESSON 4

The Big Five

Objectives

▶ Students will help to create a dichotomous key useful for identifying vertebrate animals.

▶ Students will research answers to questions they have about the five vertebrate animal groups.

Materials

▶ *The Big Five* poem from last class period

▶ Research materials for students on the five vertebrate groups

▶ One piece of poster board

▶ Marker

▶ Print out of the vertebrate dichotomous key cards

 DOI: 10.4324/9781003334743-5

Assessments

▶ *Vertebrate dichotomous key* performance task

▶ Journal prompt

Procedures

1. Review major concepts from the last class period (e.g., purpose of dichotomous keys, names of the five vertebrate animal groups).

2. Read to students the poem *The Big Five* used during the last class period once again.

3. Poll students to see how many chose each of the five vertebrate groups in response to the journal prompt from last class period, "Which do you know the least about?".

4. Ask students to share one of the questions they wrote down in response to the journal prompt from last class period, "What would you like to know about them?".

5. Provide research materials to students and challenge them to spend ten minutes looking up the answers to their questions.

6. After students find the answers to their questions, ask each student to share what they discovered and where they found their answer.

7. Explain to students that we are now going to see if we can create a dichotomous key together that will reveal to us the structure of vertebrate animal groups.

8. Recreate the blank template shown here either on the whiteboard, screen, or on a large piece of posterboard. (NOTE: Each of the rectangles should be the same size as the vertebrate dichotomous key cards included below.)

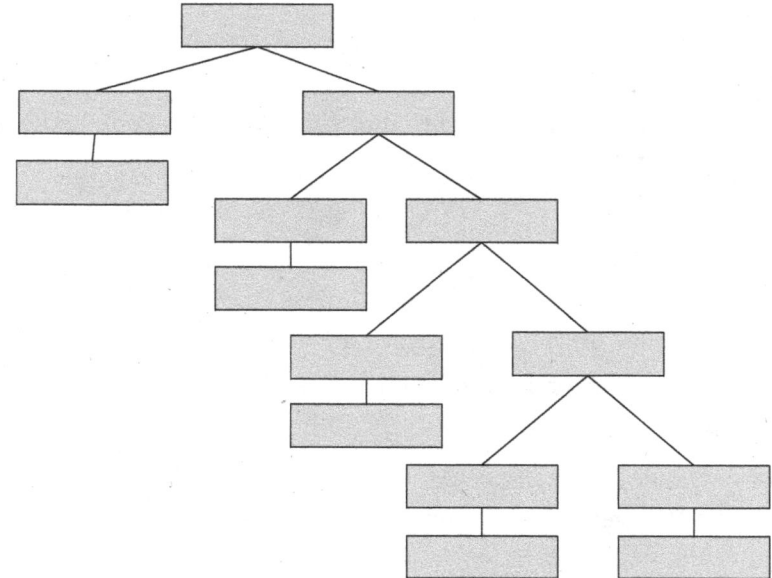

9. Print and cut out the premade dichotomous key cards included below.

10. Challenge students to work together to place the cards on the key in a way that makes sense. Remind students of the *guess and check* strategy, which means they may need to rearrange or start over several times until they find a solution that makes sense. The correct answers are shown below.

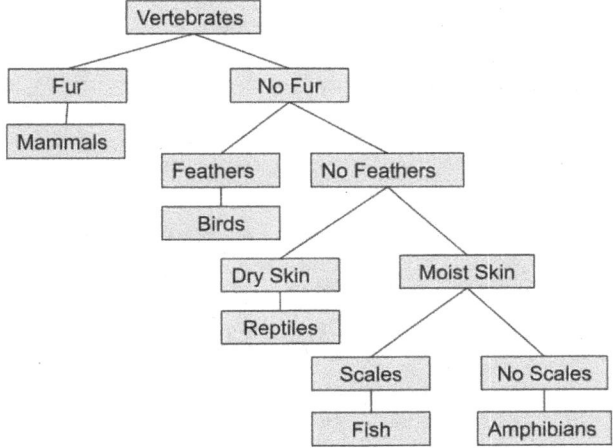

11. Once students have figured out the correct structure, name an animal for them and ask them to use the key to sort it into the correct group.

12. Challenge each student to name an animal of their own and verify the accuracy of the key by seeing if it works to sort the animal they chose.

13. Close out the lesson by asking students to complete this journal prompt: "Do you think it is possible for a vertebrate animal to fall into more than one of the big five groups? Why or why not?".

TEACHER'S NOTE

When printing, it may help your students complete this task if the descriptors are printed on one color paper and the names of the animal groups are printed on another.

Vertebrate dichotomous key cards

Vertebrates
Fur
Fur
Feathers
Feathers
Dry Skin
Moist Skin
Scales
Scales
Mammals
Birds
Reptiles
Fish
Amphibian

LESSON 5

Animals that Blur the Lines

Objectives

▶ Students will create a fictional animal that has characteristics from three of the five vertebrate animal groups.

▶ Students will research true hybrid animals that exist in nature.

Materials

▶ Dice

▶ Paper

▶ Computers for conducting research

31   DOI: 10.4324/9781003334743-6

Assessments

► Student-generated drawings
► Journal prompt

Procedures

1. Review major concepts from the last class period (e.g., vertebrate animal dichotomous key structure, etc.).

2. Ask students what the word *hybrid* means. Define this as *a mixture of two or more things*.

3. Remind students of the journal prompt they responded to in the last class period. Ask students to share their answers with a partner.

4. Explain to students that there are indeed animals that blur the lines or break certain rules for classification. Begin by displaying a picture of a green sea slug (*elysia chlorotica*). Tell students that this animal could be considered part plant and part animal because it steals genes from the algae it eats and can therefore make its own food like a plant.

5. Next, show students a picture of a platypus. Explain that this animal has the body of an otter, a tail like a beaver, a bill and feet like a duck, lays eggs, and is actually venomous. Ask students to debate if it is a mammal, a bird, or something else. Make sure they explain their thinking. Do not confirm or deny the accuracy of their thoughts until the debate has ended.

6. Reveal to students that scientists actually consider the platypus to be a mammal because it has fur and feeds its young milk.

7. Give students time to research some of the other true hybrids listed below:
   a. Pangolins – the only mammal with scales.
   b. Bats – the only mammal that truly flies.
   c. Seahorses – fish without scales that have the shape of a mammal.
   d. Caecilians (seh-CEE-lee-enz) – look like snakes or worms, but are amphibians.
   e. Tuatara – look like lizards, but are the only living species of a now extinct group called the Rhynchocephalia.

8. Explain that students will now create their own fictional hybrid animal. Give each student a blank piece of paper and ask them to roll a dice three times. The numbers they roll will determine, in part, what their animal will look like. The first roll will determine the face of the animal. The second roll will determine the body of the animal. The third roll will determine the legs or

mode of transportation for the animal. Each number on the dice corresponds to the following animal groups:

1 = choose a trait of a mammal.
2 = choose a trait of a fish.
3 = choose a trait of a reptile.
4 = choose a trait of a bird.
5 = choose a trait of an amphibian.
6 = roll again.

9. When students are finished designing their animals, lead them through a gallery walk so that they may see each of the newly created creatures.

10. Journal prompt: "We have looked at seven different animals that blur the lines today. What does this tell you about the structure of animal classification? Could it change in the future? Why or why not?".

New Zoo for You

Objectives

▶ Students will be introduced to a vertebrate animal problem-based learning task.

▶ Students will choose an animal they know little about and begin researching it.

Materials

▶ *Need to Know Board* handout

▶ Various books or online resources for researching animals

▶ *A 20 Question Creature Quest* handout

 DOI: 10.4324/9781003334743-7

Assessments

▶ *Need to Know Board*

▶ *A 20 Question Creature Quest* handout

▶ Journal prompt

Procedures

1. Review major concepts from the last class period (e.g., examples of animals that blur the lines, etc.).

2. Ask the students to raise their hand if they have ever been to a zoo. Give students a few minutes to talk to a partner about the last time they visited the zoo. What did they see, hear, eat, smell, do?

3. Introduce the problem-based learning prompt using the script below:

> Most of us find going to the zoo quite enjoyable. It is an opportunity to learn about animals we may not know much about or get to see in the wild. Zoos do a lot to preserve and protect endangered animals around the world as well. Zoos change over time just like people and other things do. The first zoos had small cages for the animals to live in instead of exhibits that looked like their real habitat. (*Show pictures below.*) Imagine you are a world-renowned zoologist and a local zoo has called you for advice on how to design a new innovative zoo exhibit for an animal of your choosing. You must choose a vertebrate animal that you know little about for this task. The zoo exhibit must include not only that animal's habitat, but an entire functioning ecosystem. Your budget is unlimited.

Old-Style Zoo Exhibit	*New-Style Zoo Exhibit*

4. Give students a copy of the *Need to Know Board* and ask them to record in column one everything they know about the problem. Examples might include such things as:

    a. We are pretending to be a famous zoologist.

    b. We have to design a zoo exhibit.

    c. The animal has to be one we know little about.

    d. It has to include this animal's entire ecosystem.

    e. Etc.

5. Next, ask students to begin listing what they need to find out in order to complete the task or solve the problem in column two. Examples might include such things as:

    a. What is an ecosystem?

    b. Which animal will I choose?

    c. Where does my animal live?

    d. What materials will I need to construct my exhibit?

    e. Etc.

6. Provide various books and online resources for students to begin searching for a vertebrate animal they would like to choose. Remind students that they must choose an animal they do NOT already know a lot about.

7. Distribute a copy of the *A 20 Question Creature Quest* handout to each student. Ask students to answer **questions 1 through 3** when they have chosen the vertebrate animal of their choice.

8. Journal prompt: "What do you think it takes to become a zoologist for a career?"

TEACHER'S NOTE

If you are unfamiliar with problem-based learning and how to adapt it for gifted students, locate and read the following resource:

Gallagher, S. A. (2009). Adapting problem-based learning for gifted students. In F. A. Karnes & S. M. Bean (Eds.), *Methods and materials for teaching the gifted* (pp. 301–330). Prufrock Press Inc.

Need to Know Board

Name _____   Date _____

What We KNOW About the Problem	What We NEED to Know	Where We Can Find Out	✓

A 20 Question Creature Quest

Name _____   Date _____

Directions: Please answer each of the 20 questions below when prompted.

WHICH VERTEBRATE ANIMAL DID YOU DECIDE TO RESEARCH?

_____

---

1. To which animal family does it belong? (circle one)
   Amphibian   Bird   Fish   Mammal   Reptile

2. List what you **already know** about your animal below:

3. What **unanswered questions** do you have about this animal?

4. In which **part of the world** does your animal live (e.g., states, countries, continents, etc.)?

5. In what **habitat** does it live?

6. What **other plants and animals (biotic organisms)** live in its habitat?

7. What **non-living (abiotic) things** does your animal interact with in its ecosystem?

8. How have humans impacted your animal's ecosystem? List any **positive indicators** or **negative stressors** you can find to back up your answer.

9. Is your animal a **carnivore, omnivore, or herbivore**? What specifically does it eat? How often does it eat?

10. Does your animal have any special adaptations that either help it **avoid becoming prey** or help it become a **better predator**? If so, what are they?

LESSON 7

11. How much **space** does the animal need to live comfortably and naturally in its habitat?

12. Is your animal **solitary or communal**? If communal, how large is a typical group?

13. What is the **collective noun** for a group of your animal (e.g., herd, flock, school, etc.)?

14. Is your animal **nocturnal** or **diurnal**?

15. Does your animal **hibernate** or **estivate**? If so, which one and when?

16. In which **temperatures** and with what amount of **moisture** does your animal thrive?

17. How **quickly** does your animal move? Is it an **endotherm** or **ectotherm**?

18. Does your animal use **concealing or disruptive coloration**? If so, how?

19. Is your animal **endangered or threatened**? If so, why is this happening?

20. How **intelligent** is your animal? Give an example of an **instinct** and a **learned behavior**?

**PLEASE USE THE SPACE BELOW TO RECORD THE ANSWERS
TO THE QUESTIONS YOU LISTED IN QUESTION 3.**

▶

▶

▶

Image Credits
Cory Seamer/Shutterstock.com
kolalok/Shutterstock.com

LESSON 7

Evaluating an Ecosystem

Objectives

- ▶ Students will distinguish between a habitat and an ecosystem.
- ▶ Students will categorize things as biotic or abiotic.

Materials

- ▶ *Need to Know Board* handout
- ▶ *I have ... Who has ...* game cards
- ▶ Ecosystem image of your choice
- ▶ Paper and pencil
- ▶ *A 20 Question Creature Quest* handout
- ▶ Various children's books or nonfiction reference materials about different habitats

 DOI: 10.4324/9781003334743-8

Assessments

▶ *Need to Know Board*

▶ *A 20 Question Creature Quest* handout

▶ Journal prompt

Procedures

1. Review major concepts from the last class period (e.g., details of the PBL task they were given, animal they chose, etc.).

2. Begin by asking the students if they know what the word *habitat* means.

3. Define the word habitat as *a place where an animal lives naturally.*

4. Play the game *I have ... Who has ...* by cutting apart the cards below. Give one card to each student. Make sure that you (a) shuffle the cards before passing them out, and (b) all cards have been distributed. Ask the **duck** to begin by reading their card. The person who can answer the duck's question will go next. Continue until the duck is called on again.

5. Explain to the students that an ecosystem is different from a habitat. Poll the students to see if anyone knows what the difference is.

6. Define the word *ecosystem* as *communities of living things interacting with the nonliving parts of their environments.*

7. Ask if anyone knows the scientific words for living and nonliving. Introduce the terms *biotic* and *abiotic*. Explain that *bio* is Greek for the word *life* and the prefix '*a*' is Greek for *not or without*.

8. Show students a beautiful picture of an ecosystem (e.g., tropical rainforest scene, a desert scene, an ocean scene, etc.).

9. Ask the students to create a T-chart on a piece of paper and label the first column *Biotic* and the second column *Abiotic*. Challenge students to record what they see under the correct heading (e.g., BIOTIC – animals, plants, etc. /ABIOTIC – sand, water, clouds, rocks, etc.).

10. Read each of the scenarios below and ask the students to determine with a partner if each one is an example of a habitat or an ecosystem:

    a. A bear hunts a deer along a riverbed in the middle of a forest (i.e., ecosystem).

    b. A snake lives in a burrow underground (i.e., habitat).

    c.   A large school of fish under the water are being chased by a dolphin. The fish accidentally startle a flock of seabirds floating nearby (i.e., ecosystem).

    d.   The gila woodpecker hollows out a home for itself inside a saguaro cactus (i.e., habitat).

11.   Challenge each pair of students to list their own example of a habitat and an ecosystem. Give students time to share what they created with the whole class.

12.   Ask students to now shift gears to thinking about the animals they are researching for their PBL projects. Ask students to get out the *A 20 Question Creature Quest* handout and answer **questions 4–7** while conducting any necessary research.

13.   Remind students to add any new things they have learned or new questions they need to find the answers to on their *Need to Know Boards*.

14.   If students finish early, provide a sampling of books about various habitats and ecosystems for students to read.

15.   Journal prompt: "How would you describe the habitat that YOU live in? What are the most important biotic things in your life? What are the most important abiotic things in your life?"

I have ... Who has ... game cards

I have ... Who has ... **I have a duck.** **Who has the habitat where ducks live?**	*I have ... Who has ...* **I have a pond.** **Who has another animal that lives in or near a pond?**
I have ... Who has ... **I have a snapping turtle.** **Who has the habitat where sea turtles live?**	*I have ... Who has ...* **I have the ocean.** **Who has another animal that lives in the ocean?**
I have ... Who has ... **I have a seahorse.** **Who has the manmade habitat where four-legged horses often live?**	*I have ... Who has ...* **I have a farm.** **Who has another animal that lives on a farm?**
I have ... Who has ... **I have a pig.** **Who has the habitat in Arizona where an animal that looks like a pig, called a javelina, lives?**	*I have ... Who has ...* **I have the desert.** **Who has another animal that lives in the desert?**

I have … Who has …

I have a burrowing owl.

Who has the habitat where a snowy owl lives part of the year?

I have … Who has …

I have the Arctic where it is cold much of the year.

Who has another animal that visits the Arctic?

I have … Who has …

I have a walrus.

Who has the habitat in Africa where another animal with long tusks, an elephant, lives?

I have … Who has …

I have the savanna.

Who has another animal that lives in the savanna?

I have … Who has …

I have a cheetah.

Who has the jungle habitat in South America where another large cat, the jaguar, lives?

I have … Who has …

I have the Amazon rainforest.

Who has another animal that lives in the rainforest?

I have … Who has …

I have a toucan.

Who has the habitat where a bird that doesn't fly, the penguin, lives?

I have … Who has …

I have Antarctica.

Who has the name of a bird that lives in freshwater ponds?

How Humans Help or Hinder Habitat Health

Objectives

▶ Students will design a public park that incorporates three or more habitats.

▶ Students will investigate the impacts humans have on the environment.

Materials

▶ Large piece of butcher paper

▶ Crayons and markers

▶ Various pieces of clean trash

▶ Tape

▶ Book *Wartville Wizard* by Don Madden

 DOI: 10.4324/9781003334743-9

▶ *Need to Know Board* handout

▶ *A 20 Question Creature Quest* handout

Assessments

▶ *Need to Know Board*

▶ *A 20 Question Creature Quest* handout

▶ Journal prompt

Procedures

1. Review major concepts from the last class period (e.g., habitat versus ecosystem, biotic versus abiotic, etc.).

2. Ask the students to talk with a partner and determine if it is possible to have (a) more than one ecosystem in a habitat, (b) more than one habitat in an ecosystem, or (c) if neither is possible.

3. Explain that the correct answer is (b). You could have more than one habitat in an ecosystem.

4. To prove your point, ask students to think of a beautiful public park. Explain that, as an example, you could have a freshwater stream, a woodland area, and a grassland area within a small space.

5. Challenge students to create a class mural showing the perfect public park. Provide one large piece of butcher paper, crayons, and markers. Allow students to work for 10–15 minutes.

6. Hang the park mural on the wall. Count the number of habitats together.

7. Next, without saying a word, begin taping clean trash (e.g., paper cup, plastic bag, ketchup packet, straw, etc.) all over the finished piece of art.

8. Listen to the students' reaction and ask how it makes them feel.

9. Explain that littering and improper trash disposal is a big problem all over the world.

10. Read the book *Wartville Wizard* by Don Madden. (*Summary*: a man is tired of people littering all over his yard and is magically given power over trash. Now, whenever anyone litters on his property, he just points to the trash and it flies back to the person who threw it down and sticks to their body.)

    Debrief the book and show a few real pictures of human impact on the environment. A few examples are included here:

11. Ask how we as a school can positively impact the environment where we live. Discuss together or have students brainstorm ideas with a partner.

TEACHER'S NOTE

This step should not be skipped. Many gifted students are sensitive about world issues such as human pollution, climate change, etc., and feel helpless to make a difference. This can lead to what is known as existential depression. Giving students the opportunity to brainstorm a solution and make a difference will relieve much of the stress they may feel.

12. Ask students to answer **part one** of **question 8** on *A 20 Question Creature Quest* handout after conducting any necessary research.

13. Remind students to record any new information or questions on their *Need to Know Boards*.

14. Journal prompt: "I think it is important to protect the environment because … "

Image Credits
Bystrov/Shutterstock.com
hxdbzxy/Shutterstock.com
VanderWolf Images/Shutterstock.com

An Ecosystem Stress Test

Objectives

- ▶ Students will discuss ways to determine the health of an ecosystem.
- ▶ Students will test the health of a fictional ecosystem using a simulation/ game.

Materials

- ▶ *Equilibrium Stress Test* handout
- ▶ *Equilibrium Stress Test Data Collection Tool* handout
- ▶ One penny or dime for each student
- ▶ One dice for each student
- ▶ *Need to Know Board* handout
- ▶ *A 20 Question Creature Quest* handout

51   DOI: 10.4324/9781003334743-10

Assessments

▶ *Need to Know Board*

▶ *A 20 Question Creature Quest* handout

▶ Journal prompt

Procedures

1. Review major concepts from the last class period (e.g., how humans can negatively impact the environment, etc.).

2. Explain to students that there are ways to tell how healthy an ecosystem actually is. Distribute the *Ecosystem Stress Test* handout and discuss the six examples listed under *Positive Indicators* on the left side of the scale.

3. Tell students that we call things that are negatively impacting an ecosystem *stressors*. Read through and discuss the *Negative Stressors* on the right side of the scale.

4. Challenge students to list any additional positive indicators or negative stressors they can think of that could have been included on the handout.

5. Give each student a dice and a penny or dime. Explain that they will use these to test the health of a pretend ecosystem while following the directions stated on the handout.

6. Provide time for students to play the game several times.

7. Debrief the activity and ask students to draw conclusions from what they saw (e.g., it often doesn't take one major stressor to cause an ecosystem to collapse, but many small stressors instead).

8. Ask students to think about the ecosystem in which they live. How would they rank the health of it using the diagnostic scale at the bottom of the handout? Make sure students justify their answers with data or observations.

9. Ask students to research and answer **part two** of **question 8** on *A 20 Question Creature Quest* handout.

10. Ask students to record any new information on their *Need to Know Boards*.

11. Journal prompt: "Compare your answer to number 8 on *A 20 Question Creature Quest* handout with those of your classmates. What did you find similar and different?"

TEACHER'S NOTE

Many students are unfamiliar with lichen. Consider showing a picture or bringing in a sample from your school yard or a local park. Ask the students if they think lichens are a plant, fungus, or something else. In actuality, lichens are made of a fungi living symbiotically with algae. There are over 3,600 different varieties of lichen around the world.

Ecosystem Stress Test

Name _____   Date _____

Directions: Find a coin and one dice. Flip the coin. If it lands on HEADS, roll the dice and add points to your starting total of 200 according to the chart under POSITIVE INDICATORS. If it lands on TAILS, roll the dice and subtract points from your starting total of 200 according to the chart under NEGATIVE STRESSORS. Repeat the process nine more times. Compare your final total with your original total of 200. If the total is less than 110, your ecosystem is not healthy. If it is between 110–160, your ecosystem is at risk. If it is 161–240, your ecosystem is healthy. Above 240 means your ecosystem is extremely healthy.

POSITIVE INDICATORS	
Rolled a 1	Frog and dragonfly populations strong (+20 points)
Rolled a 2	Presence of lichen found (+15 points)
Rolled a 3	No invasive species found (+25 points)
Rolled a 4	Bee populations are rebounding (+10 points)
Rolled a 5	Local farmers decide to go organic (+15 points)
Rolled a 6	Temperature and precipitation levels remain stable (+30 points)

NEGATIVE STRESSORS	
Rolled a 1	Natural disaster hits (-10 points)
Rolled a 2	New city being built (-40 points)
Rolled a 3	Deer population explodes (-10 points)
Rolled a 4	Invasive species found (-15 points)
Rolled a 5	New fungus begins killing off some species of trees (-20 points)
Rolled a 6	Water quality so poor that species such as herons move away (-35 points)

EQUILIBRIUM

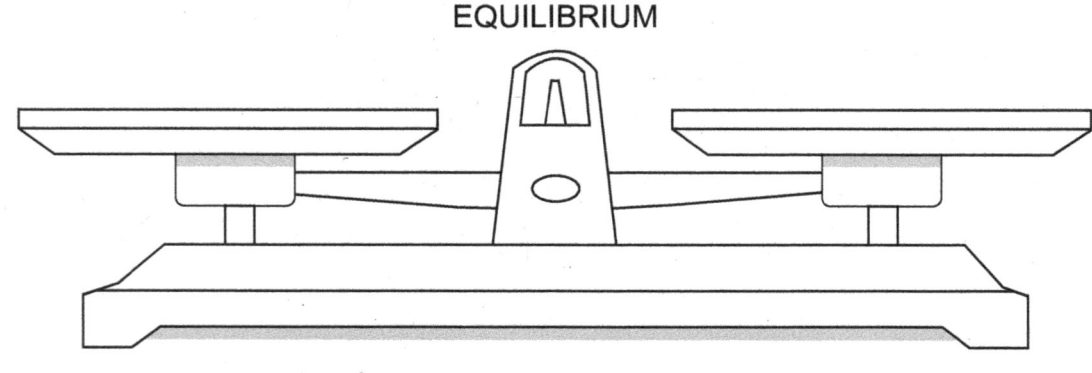

Ecosystem Stress Test Data Collection Tool

Name _____   Date _____

Directions: Record your new total after each turn you take. Analyze the health status after ten rolls.

	Game 1	Game 2	Game 3	Game 4	Game 5
Beginning Total	200	200	200	200	200
Total After Roll 1					
Total After Roll 2					
Total After Roll 3					
Total After Roll 4					
Total After Roll 5					
Total After Roll 6					
Total After Roll 7					
Total After Roll 8					
Total After Roll 9					
Total After Roll 10	Final Total:	Final Total:	Final Total:	Final Total:	Final Total:
Health Status					

Diagnostic Scale

NOT Healthy / Declining	**Below 110**
At Risk / Unstable	**110–160**
Healthy / Stable	**161–240**
Extremely Healthy	**Over 240**

LESSON 10

Predator or Prey Predictions

Objectives

▶ Students will classify animals as being either predator or prey.

▶ Students will examine adaptations animals have that help them avoid becoming prey or become a better predator.

Materials

▶ Pictures of animal skulls

▶ *Predator or Prey* cards

▶ *What Do You Do When Something Wants to Eat You* by Steve Jenkins

▶ Tarp or blanket

▶ Several cups of dry cereal

 DOI: 10.4324/9781003334743-11

▶ *Need to Know Board* handout

▶ *A 20 Question Creature Quest* handout

Assessments

▶ *Need to Know Board*

▶ *A 20 Question Creature Quest* handout

▶ Journal prompt

Procedures

1. Review major concepts from the last class period (e.g., positive indicators and negative stressors on the environment, etc.).

2. Explain to students that another way to classify animals is according to what they eat. Ask the students what you call something that eats meat (i.e., carnivore), versus something that eats plants (i.e., herbivore) or both meat and plants (i.e., omnivore).

3. Tell students that one way to determine what something eats is to look at its teeth. Show students the two pictures of animal skulls found below. Ask them what type of food they think each creature ate and why.

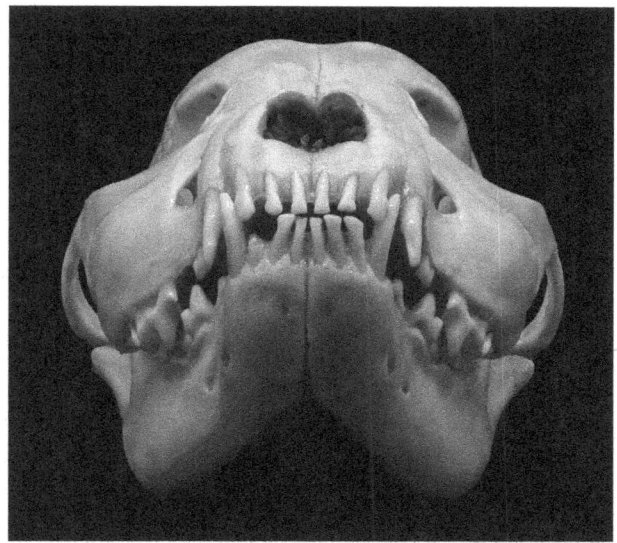

4. If time permits, show students additional skulls and ask them to guess (a) what kind of animal it is, and (b) if it is a carnivore, herbivore, or omnivore.

5. Explain that carnivores and omnivores are considered *predators* of the animals they eat, but the animal that gets eaten is called the *prey*.

6. Ask the students what the difference is between PRAY and PREY. Poll the students to see if anyone knows what you call two words that sound the same, but have different meanings (i.e., homophones).

7. Give each student a *Predator or Prey* animal card and ask them to research if the animal is a predator or prey. If it is clearly prey, list one of its predators. If it is clearly a predator, list one of its prey. If it depends on the situation (e.g. a cat pursuing a mouse, versus a coyote pursuing a cat), ask the students to give an example of when that animal could be both. Allow students to collaborate with a partner if they do not know enough about the animal to make a determination.

8. Explain that animals have adapted over the centuries in order to survive. Read the book *What Do You Do When Something Wants to Eat You?* by Steve Jenkins.

9. Ask students if they know of any other animals who have special traits that allow them to escape predators.

10. If time permits, play the game called *Wolf Pack* described below outside:

> Lay a plastic tarp on the ground and sprinkle a cup or two of dry cereal on top. Divide the students into two halves. Group one will be deer and group two will be wolves. The deer will begin by being on their knees on the tarp gathering the cereal for food. The wolves will begin approximately 75 to 100 feet

away and must slowly creep towards the deer. At any point, the wolves may chase the deer and attempt to tag them. If they are tagged, they have become prey and are out of the game. If the deer are not tagged and make it to a designated safe zone, then the wolves go hungry and the deer live on. Switch the groups and repeat.

11. Debrief the simulation and ask students to share how it felt to be both predator and prey.

12. Ask students to answer **questions 9 and 10** on *A 20 Question Creature Quest* handout for their chosen animal.

13. Remind students to also add to their *Need to Know Boards*.

14. Journal prompt: "What would you do if something wanted to eat you?"

TEACHER'S NOTE

Consider connecting this lesson with what you are teaching in language arts or grammar. Students could brainstorm additional homophones, create a class dictionary of homophones in which they illustrate and label homophone pairs, discuss the differences between homophones and homonyms, etc.

Predator or Prey cards

PREDATOR or PREY? LION	**PREDATOR or PREY?** BEAVER	**PREDATOR or PREY?** SQUIRREL	**PREDATOR or PREY?** SALMON
PREDATOR or PREY? RABBIT	**PREDATOR or PREY?** RAVEN	**PREDATOR or PREY?** GECKO	**PREDATOR or PREY?** BEAR
PREDATOR or PREY? PANDA	**PREDATOR or PREY?** GORILLA	**PREDATOR or PREY?** DONKEY	**PREDATOR or PREY?** ALLIGATOR
PREDATOR or PREY? FROG	**PREDATOR or PREY?** SNAKE	**PREDATOR or PREY?** OSTRICH	**PREDATOR or PREY?** GIRAFFE
PREDATOR or PREY? ELEPHANT	**PREDATOR or PREY?** FOX	**PREDATOR or PREY?** PARROT	**PREDATOR or PREY?** CLOWN FISH
PREDATOR or PREY? CHEETAH	**PREDATOR or PREY?** SLOTH	**PREDATOR or PREY?** DOLPHIN	**PREDATOR or PREY?** MANATEE

Image Credits

Bruce MacQueen/Shutterstock.com

Fercast/Shutterstock.com

Elbow Room

Objectives

▶ Students will determine the amount of space various animals require.

▶ Students will classify animals as either solitary or communal.

Materials

▶ One set of *Species Cards* and *Collective Noun Cards* per small group

▶ *Need to Know Board* handout

▶ *A 20 Question Creature Quest* handout

 DOI: 10.4324/9781003334743-12

Assessments

- *Need to Know Board*
- *A 20 Question Creature Quest* handout
- Journal prompt

Procedures

1. Review major concepts from the last class period (e.g., predator, prey, carnivore, omnivore, herbivore, etc.).

2. Explain to students that some people are *introverts* (i.e., prefer to be alone and get energy from being alone) and some people are *extroverts* (i.e., prefer to be with others and get energy from being around others).

3. Poll the students to see which category they personally believe they most fit. Stress to students that neither is right or wrong.

4. Tell students that in the animal kingdom the same is true. Some animals are *solitary* (i.e., prefer living alone) and others are *communal* (i.e., prefer living in groups).

5. Group students into pairs and ask them to come up with two or three animals that fall into each category. A few examples of solitary creatures are wolverines, skunks, and snow leopards. A few examples of communal animals are elephants, wolves, and dolphins.

6. Provide an opportunity for students to share.

7. Challenge students to complete the following analogy:
   A bird is to a flock, as a kitten is to a _____.

8. Explain to students that words that name a group of animals are called *collective nouns*. Some collective nouns are very common (e.g., flock, pack, herd), but others are not (e.g., murder of crows, mess of iguanas, etc.).

9. Tell students that those names were chosen long ago, most likely because of a trait those animals possess. Ask students to imagine a group of iguanas lying in the sun. It would look like a jumbled mess of tails and heads.

10. Challenge students to match each animal species on the *Species Cards* included below with the correct collective noun on the *Collective Noun Cards*. Be sure to shuffle before giving them to students.

11. After students have made their pairings, reveal the correct answers shown in the chart below.

Species	Collective Noun	Species	Collective Noun
Bat	Colony	Jellyfish	Smack
Bee	Swarm	Hippopotamus	Bloat
Camel	Caravan	Lion	Pride
Goose	Gaggle	Hyena	Cackle
Gorilla	Troop	Owl	Parliament
Fish	School	Otter	Family or romp
Frog	Army	Kangaroo	Mob
Giraffe	Tower	Mule	Pack
Parrot	Pandemonium	Shark	Shiver
Porcupine	Prickle	Cow	Herd
Rabbit	Colony or fluffle	Pig	Drift or team
Stingray	Fever	Rat	Colony or mischief
Squirrel	Scurry	Lemur	Conspiracy
Turtle	Bale	Flamingo	Flamboyance
Whale	Pod	Cobra	Quiver
Zebra	Dazzle	Crocodile	Bask

12. Tell students that in addition to being solitary or communal, some animals need a lot of space to roam (i.e., elbow room) while others do not. List three example types of animals that need a lot of space listed below:

    a. Apex predators need room to hunt.

    b. Pollinators need a large area to find plants.

    c. Migrating animals need places to travel to when conditions or seasons change.

13. Ask students to think of examples of animals that are content with living in small spaces. Explain that, in general, these animals tend to be smaller than those which need a lot of space to roam.

14. Instruct students to answer **questions 11 through 13** on *A 20 Question Creature Quest* handout.

15. Remind students to also add to their *Need to Know Boards*.

16. Journal prompt: "How much 'elbow room' would you need if YOU were an animal?"

TEACHER'S NOTE

Taking time to help students understand who they are and how they learn best is a valuable tool for both you and the students. You can use the information to help differentiate instruction and the students can use the information to advocate for themselves and find others like themselves. Consider giving students a kid-friendly introvert/extrovert assessment and take note on how each student identifies.

Species Cards

Species of Animal **Bat**	Species of Animal **Jellyfish**
Species of Animal **Bee**	Species of Animal **Hippopotamus**
Species of Animal **Camel**	Species of Animal **Lion**
Species of Animal **Goose**	Species of Animal **Hyena**
Species of Animal **Gorilla**	Species of Animal **Owl**
Species of Animal **Fish**	Species of Animal **Otter**
Species of Animal **Frog**	Species of Animal **Kangaroo**
Species of Animal **Giraffe**	Species of Animal **Mule**
Species of Animal **Parrot**	Species of Animal **Shark**
Species of Animal **Porcupine**	Species of Animal **Cow**
Species of Animal **Rabbit**	Species of Animal **Pig**
Species of Animal **Stingray**	Species of Animal **Rat**
Species of Animal **Squirrel**	Species of Animal **Lemur**
Species of Animal **Turtle**	Species of Animal **Flamingo**
Species of Animal **Whale**	Species of Animal **Cobra**
Species of Animal **Zebra**	Species of Animal **Crocodile**

Collective Noun Cards

Collective Noun	Collective Noun
Colony	**Smack**
Collective Noun **Swarm**	Collective Noun **Bloat**
Collective Noun **Caravan**	Collective Noun **Pride**
Collective Noun **Gaggle**	Collective Noun **Cackle**
Collective Noun **Troop**	Collective Noun **Parliament**
Collective Noun **School**	Collective Noun **Family** or **Romp**
Collective Noun **Army**	Collective Noun **Mob**
Collective Noun **Tower**	Collective Noun **Pack**
Collective Noun **Pandemonium**	Collective Noun **Shiver**
Collective Noun **Prickle**	Collective Noun **Herd**
Collective Noun **Colony** or **Fluffle**	Collective Noun **Drift** or **Team**
Collective Noun **Fever**	Collective Noun **Colony** or **Mischief**
Collective Noun **Scurry**	Collective Noun **Flamboyance** or **Stand**
Collective Noun **Bale**	Collective Noun **Conspiracy**
Collective Noun **Pod**	Collective Noun **Quiver**
Collective Noun **Dazzle**	Collective Noun **Bask**

It's Time for Bed

Objectives

► Students will examine the sleep patterns of various animals.
► Students will be able to explain the differences between hibernate and estivate.

Materials

► *Need to Know Board* handout
► *A 20 Question Creature Quest* handout

 DOI: 10.4324/9781003334743-13

Assessments

▶ *Need to Know Board*

▶ *A 20 Question Creature Quest* handout

▶ Journal prompt

Procedures

1. Review major concepts from the last class period (e.g., communal, solitary, introvert, extrovert, etc.).

2. Pose the question *How important is sleep?* to the students and listen to their responses.

3. Explain to students that sleep is the time when both humans and animals heal, recuperate, and recharge. Introduce the terms *nocturnal* (i.e., sleep during the day) and *diurnal* (i.e., sleep during the night).

4. Challenge students to work with a partner to list five to ten of each type of animal. A few examples of animals that are nocturnal are owls, sugar gliders, and aardvarks.

5. Ask students to share their final lists.

6. Pose this question to students – *Zoos are typically not open during the night. How might a zoo TRICK a nocturnal animal into thinking it is night during the day so that it is up and moving for the zoo exhibitors?*

7. Give students time to research and share.

8. Explain to students that sleep also helps animals conserve energy and survive long winters or extreme heat causing a drought. Introduce the terms *hibernate* (i.e., sleep when it gets cold) and *estivate* (i.e., sleep when it gets too hot or dry).

9. Give students several examples of animals that estivate (e.g., African bullfrog, cane toad, lungfish, and hedgehogs). Ask students to conduct research in order to find at least five more animals that estivate.

10. Ask students to share their final lists.

11. Pose this question to students – *Zoos can be found all over the world. How do zoos help animals deal with hot or cold temperatures they are not used to in the wild?*

12. Give students time to research and share.

13. Ask students to research and record the answers to **questions 14 through 16** on the *A 20 Question Creature Quest* handout.

14. Remind students to add new information to their *Need to Know Boards*.

15. Journal prompt: "*Rip Van Winkle* is a fictional story by Washington Irving about a man who went to sleep for 20 years. When he woke up everything around him had changed. Imagine you are an animal living in a large forest that accidently hibernated for 20 years. When you wake up, you find yourself living in a tiny park surrounded by a huge bustling city. What would you think? How would you react? What would you need to do differently in your daily life?"

Speed Racer or Snail Pacer

Objectives

▶ Students will explore the different ways animals move.

▶ Students will compare and contrast the advantages of moving quickly versus moving slowly in the animal kingdom.

Materials

▶ *If You Hopped Like a Frog* by David Schwartz

▶ *Need to Know Board* handout

▶ *A 20 Question Creature Quest* handout

DOI: 10.4324/9781003334743-14

Assessments

- *Need to Know Board*
- *A 20 Question Creature Quest* handout
- Journal prompt

Procedures

1. Review major concepts from the last class period (e.g., nocturnal, diurnal, hibernate, estivate).

2. Remind students that we have been discussing the different ways animals have adapted to their environments in order to survive. Today we are going to continue that discussion by talking about how different animals move and at what speed.

3. Take students outside to play the following game:

   "Animal Relay Race"
   Divide students into two teams and create a starting line and finish line using cones or boundaries of some kind. Explain that both teams will compete against each other in a relay race. Instead of running, however, the teacher will call out an animal and the students must move like that animal. A few examples are (a) hop like a kangaroo, (b) crab walk like a crab, (c) gallop like a horse (but on two legs), or (d) fly like a bird (flapping arms), etc.

4. Bring students back inside and read the book *If You Hopped Like a Frog* by Davis Schwartz.

5. Ask the students to talk to a partner about the animal fact that surprised them the most.

6. Partner students together and ask them to name the slowest moving animal and the fastest moving animal they can think of.

7. Ask students: *What are the advantages of moving quickly*? Example answers are catch prey easier, escape from predators quicker, etc.

8. Pose the questions: *Are there any advantages to moving slowly? If so, what are they?*

9. Listen to student responses and explain that there are indeed advantages to moving slowly (e.g., keeps the animal from being spotted, allows the animal to hide until its prey comes near, etc.).

10. List some of the slowest vertebrate animals and ask students to make a generalization about them (e.g., sloth, ball python, bearded dragon, Russian tortoise, veiled chameleon, four-toed hedgehog, horned frog, etc.).

11. Explain to students that more reptiles and amphibians move slowly than mammals. Ask students to predict why that is.

12. Introduce the terms *endotherms* (use food energy to regulate body temperature) and *ectotherms* (i.e., must rely on the sun to provide heat). Ask students which of the two new terms is another name for warm blooded and which is cold blooded (i.e., endotherms are warm blooded).

13. Pose this question and ask students to discuss with a partner: *Is a cold-blooded animal's blood always cold?* Explain that the answer is no. A cold-blooded animal in the sun may have hotter blood than a warm-blooded animal in the same location.

14. Ask students how warm-blooded creatures cool down (e.g., sweat, pant, cover themselves in mud, etc.) as compared to how cold-blooded animals cool down (e.g., go underground, get in the water, etc.).

15. Ask students to research and answer **question 17** on *A 20 Question Creature Quest* handout.

16. Remind students to add new information to their *Need to Know Boards*.

17. Journal prompt: "On average, a sloth travels 41 yards a day. How many days would it take a sloth to travel from your state's capital to the capital of a state that borders your state?"

TEACHER'S NOTE

The fastest animal is the cheetah. It can go from 0 to 60 miles an hour in just a few seconds. Gifted expert Stephanie Tolan wrote a wonderful short story on her website called *Is It a Cheetah?* (www.stephanietolan.com/is_it_a_cheetah.htm). Consider reading this to your students and discussing the pressures they may feel to "run like a cheetah" all of the time or else risk the chance they will not be looked at as a "cheetah" any longer.

Show-Off or Wallflower

Objectives

- ▶ Students will differentiate between two major types of camouflage (i.e., concealing coloration and disruptive coloration).
- ▶ Students will find scientific inaccuracies in a piece of published text.
- ▶ Students will design a fictional animal adapted to live in our classroom.

Materials

- ▶ Box of colored toothpicks
- ▶ *Chameleons are Cool* by Martin Jenkins
- ▶ The poem called *The Chameleon* from the book *Zoo Doings* by Jack Prelutsky
- ▶ Various colors of construction paper

77

DOI: 10.4324/9781003334743-15

- Scissors
- Glue
- Markers
- *Need to Know Board* handout
- *A 20 Question Creature Quest* handout

Assessments

- *Need to Know Board*
- *A 20 Question Creature Quest* handout
- Journal prompt

Procedures

1. Review major concepts from the last class period (e.g., endotherm, ectotherm, etc.).

2. Remind students that we have been discussing the different ways animals have adapted to their environments in order to survive. Explain that today we are going to discuss camouflage.

3. Walk students outside into a grassy area and sprinkle a box of colored toothpicks on the ground. Challenge the students to find each one.

4. When back inside the classroom, ask students which color of toothpick was the easiest to find and which was the hardest to find.

5. Challenge students to list as many animals as they can think of that use camouflage to survive.

6. Ask the students to raise their hand if they have a chameleon on their list. Read the poem *The Chameleon* by Jack Prelutsky, then ask students if anything in the poem is inaccurate.

7. Next, read the book *Chameleons are Cool* by Martin Jenkins.

8. Ask students to identify the differences between the poem and the book (i.e., *The poem insinuates that chameleons change in order to blend in. In fact, they change color based on their emotions, temperature, etc.*).

9. Ask students what is needed to determine which version is correct (e.g., a third piece of data, references or citations, consider the background of both authors, etc.).

10. Challenge students to determine which is correct.

TEACHER'S NOTE

One way of describing this type of thinking is *evaluative thinking*. This means to closely analyze the details of something in order to make a judgement regarding its correctness. Consider providing additional practice with evaluative thinking and/or asking students when they believe this skill will be needed in their own lives.

11. Ask the students to raise their hands if they have a zebra on their list. Chances are they will not.

12. Write the words *concealing coloration* and *disruptive coloration* on the board. Poll the students to see if anyone knows what the words *conceal* and *disrupt* mean. Define *concealing coloration* as *camouflage to blend in and hide*. Define *disruptive coloration* as *camouflage to confuse or disorient a predator*.

13. Ask the students to which group they believe a zebra belongs. (Zebras use disruptive coloration and stay in large groups to confuse their attackers.)

14. Lastly, introduce the term *countershading*. Define *countershading* as *when an animal's back is dark and its underside is light*. Challenge students to think of an animal that has this adaptation.

15. Explain that many ocean animals use countershading (e.g., species of sharks, dolphins, whales, fish, penguins, etc.). Show a picture of a penguin and ask the students if they have ever thought about why it has that tuxedo look. When swimming in the water, animals below it look up and see its white ventral side (belly). This blends in with the sun streaming through the surface of the water. Animals above it look down and see its dark dorsal side (back), which blends into the deep dark ocean that stretches below it.

16. Explain that you are now going to challenge them to create a fictional animal of their own using construction paper, glue, markers, and scissors. This animal must be adapted to live **specifically in our classroom**. Give students the opportunity to brainstorm what traits they might include in their animal's design (e.g. all white to blend in with the white board, has suction cups on its feet so it can live on the ceiling in order to avoid getting stepped on by students) before passing out the materials.

17. Give students time to create their fictional classroom animal and then secretly place them in their "natural habitat".

18. Challenge each other to find all the camouflaged animals around the room.

19. Ask students if the animal they are researching for their project uses either concealing coloration (like countershading) or disruptive coloration. If so, ask them to record this information on **question 18** of *A 20 Question Creature Quest* handout.

20. Journal prompt: "If you had to choose between concealing coloration or disruptive coloration for yourself, which would you choose and why?".

Lifespan Lineup

Objectives

▶ Students will compare and contrast the typical lifespans of various animals.
▶ Students will research if their animal is threatened or endangered.

Materials

▶ *Lifespan Line Up* handout
▶ Websites linked in lesson plan
▶ *Need to Know Board* handout
▶ *A 20 Question Creature Quest* handout

 DOI: 10.4324/9781003334743-16

Assessments

▶ *Lifespan Line Up* handout
▶ *Need to Know Board*
▶ *A 20 Question Creature Quest* handout
▶ Journal prompt

Procedures

1. Review major concepts from the last class period (e.g., concealing or disruptive coloration, countershading, etc.).

2. Ask the students if they know the average lifespan of a human. In the United States, that answer is 79 years.

3. Explain to students that some animals will naturally live longer than humans (e.g., parrots) and others will live much shorter lives (e.g., dogs).

4. Ask students to complete the *Lifespan Line Up* handout below. In this activity, students are tasked with finding several animals that naturally live a certain number of years. Students may use computers or books from the classroom.

5. Write the words *endangered* and *threatened* on the board. Define *endangered* as *an animal or plant seriously at risk of extinction*. Define *threatened* as *a plant or animal likely to become endangered in the near future*.

6. Discuss the origins of the Endangered Species Act of 1973 in the United States. Explain that this law is one of the strictest environmental protection policies in the world.

7. An animal or plant can be listed as endangered in the United States if any of the following five factors are seen:

   a. present or threatened destruction, modification, or curtailment of its habitat or range,

   b. overutilization for commercial, recreational, scientific, or educational purposes,

   c. disease or predation,

   d. inadequacy of existing regulatory mechanisms, and/or,

   e. other natural or human-made factors affecting its continued existence.
   (Language from The Endangered Species Act of 1973)

8. Two organizations in the United States are responsible for determining which animals and plants to list. The first is the US Fish and Wildlife Service (FWS). Allow students to navigate to the link below to view how FWS manages this process: www.fws.gov/program/listing-and-classification.

9. The second organization is the National Oceanic and Atmospheric Administration (NOAA). Students can view the marine animals they protect at the link below:

   www.fisheries.noaa.gov/species-directory/threatened-endangered.

10. Challenge students to find and research three endangered species in their state. One resource for finding this out is the curated list below provided by FWS below:

    https://ecos.fws.gov/ecp/report/species-listings-by-state-totals?statusCategory=Listed.

11. Provide time for students to determine if the animal they are researching is threatened or endangered. If the answer is yes, students should research and record the reasons why on **question 19** of the *A 20 Question Creature Quest* handout.

12. Remind students to record any new information on their *Need to Know Boards*.

13. Journal prompt: "What is one thing you can do to help endangered animals in your state?".

Lifespan Line Up

Name _____   Date _____

Directions: This line represents the lifespan of living organisms. It begins with 0 and goes to 150 years. Your task is to find an animal that has an average lifespan that would fit on the line between each number. When you have done this, please record the name of the animal in the box on each line and its average lifespan in the circle. Please see the human example below. Lastly, please answer the two questions at the bottom of the page.

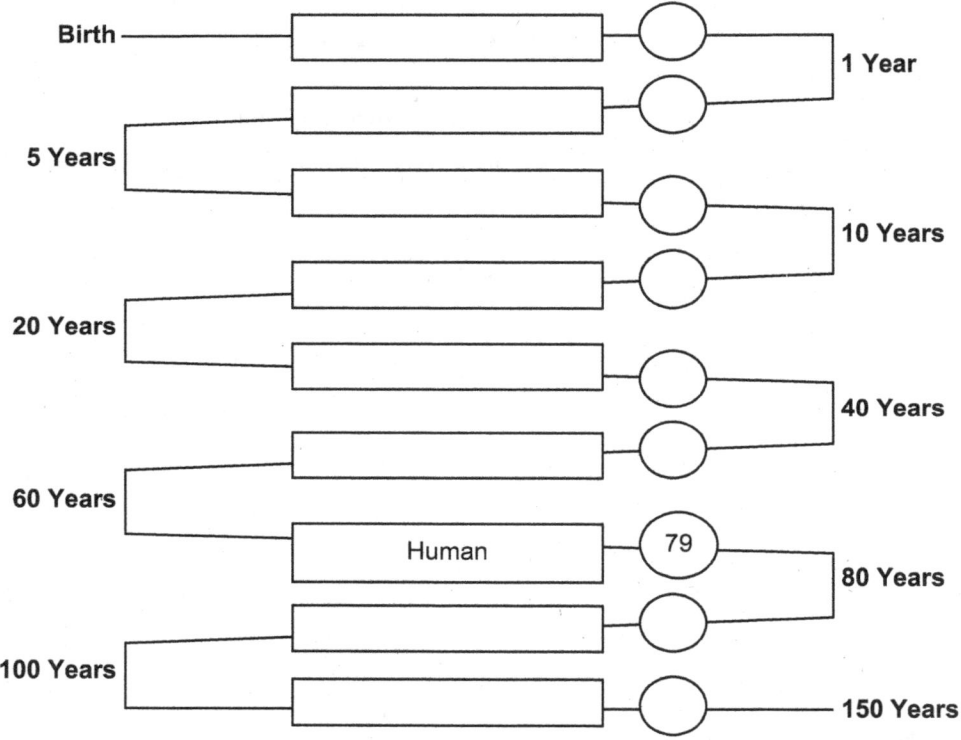

Reflection Questions

1. Which animal surprised you the most? Why?

2. What patterns or trends did you notice?

LESSON 16

Are You Smarter Than a ...

Objectives

- Students will distinguish between an instinct and a learned behavior in animals.
- Students will research signs of intelligence in the animal they chose for their project.

Materials

- *Instinct or Learned Behavior* handout
- *Are You Smarter Than a _____???* handout
- *Need to Know Board* handout
- *A 20 Question Creature Quest* handout

  DOI: 10.4324/9781003334743-17

Assessments

▶ *Instinct or Learned Behavior* handout

▶ *Are You Smarter Than a _____???* handout

▶ *Need to Know Board*

▶ *A 20 Question Creature Quest* handout

▶ Journal prompt

Procedures

1. Review major concepts from the last class period (e.g., endangered species, threatened species, etc.).

2. Ask the students to discuss with a partner the following question, "Do animals think?".

3. Provide an opportunity for students to share what they talked about with their partners.

4. Explain to students that in the past, people believed that animals were incapable of intelligence, emotions, relationships, etc. Now, however, we know differently.

5. Share the two examples below with students:

   a. An African gray parrot named Alex proved to the world that birds are not "bird-brained" or unintelligent. Alex learned over 100 words and could recognize different shapes and colors. His last words to his owner when he passed away were, "You be good, see you tomorrow. I love you".
   National Geographic. (2017). Inside Animal Minds:
   What they think, feel, and know, p. 10

   b. An octopus named Inky was held in captivity at an aquarium in New Zealand. One night, after everyone had gone home, he escaped his tank, crawled across the floor, and went down a 164-foot drainpipe. In the end, he found his way back to the ocean.
   National Geographic. (2017). Inside Animal Minds:
   What they think, feel, and know, p. 34

6. Write the word *instinct* on the board and ask if any student knows what it means. Define *instinct* as *an action that happens automatically without being taught*.

7. Distribute the handout *Instinct or Learned Behavior* and complete the first example together. Challenge the students to complete the rest of the scenarios on their own.

8. Go over the correct answers as a group and then see if students can think of any other examples of this type of behavior independently.

9. Explain to students that they will spend the rest of class researching how smart the animal they chose for their project actually is. Ask students if they have ever heard of the game show *Are You Smarter Than a 5th Grader?* Explain the premise of the show (i.e., adults compete against a panel of fifth graders to see who is smarter).

10. Distribute the handout *Are You Smarter Than a _____???* and ask students to write the name of their animal on the line.

11. Read through the *signs of intelligence* that are listed in the chart. Provide the remainder of class for students to research examples of their animal ever exhibiting any of these traits.

12. Ask students to write a summary of their findings on **question 20** of the *A 20 Question Creature Quest* handout.

13. Remind students to add any new information to their *Need to Know Boards*.

14. Journal prompt: "If you could have a real conversation with any animal, which animal would you choose? List three questions you would ask".

TEACHER'S NOTE

Intelligence is something gifted educators seek to develop and foster. Most would agree, however, that things like humility, kindness, community, creativity, etc., are also equally important. In this lesson, students were asked to compare the intelligence of one species with another. Make sure to stress to students that this should not extend to comparing or judging the intelligence of one person against the intelligence of another. Everyone has value and is special in their own way.

"The only person you should try to be better than is who you were yesterday".
– Unknown

Instinct or Learned Behavior

Name _____ Date _____

Directions: Examine each of the scenarios below and determine if it is instinct (e.g., genetic, inborn, a reflex, etc.), or if it is a learned behavior (e.g., mimicry, trial and error, etc.). Be sure to explain why you chose the choice you did.

(1)

Triangles A, B, and C watch the parent triangle in order to see how they should react.
Which is it? *(Please circle one)*
INSTINCT   or   LEARNED BEHAVIOR

Parent

Why did you circle the answer above?

Real animal that behaves in this way →	

(2)

Before Being Touched

After Being Touched

The segmented rectangle was moving silently along until it was touched. After that it curled into a ball.
Which is it? *(Please circle one)*
INSTINCT   or   LEARNED BEHAVIOR

Why did you circle the answer above?

Real animal that behaves in this way →	

LESSON 17

(3)

The parent square notices one of her eggs has fallen out of the nest. She goes to retrieve it.

_____

Which is it? *(Please circle one)*

INSTINCT   or   LEARNED
                    BEHAVIOR

Why did you circle the answer above?

Real animal that behaves
in this way →

(4)

When the bell rings, the hexagons come running for their dinner.

_____

Which is it? *(Please circle one)*

INSTINCT   or   LEARNED
                          BEHAVIOR

Why did you circle the answer above?

Real animal that behaves
in this way →

(5) Create your own INSTINCT scenario using STARS and your own LEARNED BEHAVIOR scenario using HEARTS.

Image Credits

Perfect_kebab/Shutterstock.com
Rasuke desain/Shutterstock.com

LESSON 17

Are You Smarter Than a _____ ???

Name _____   Date _____

Directions: As you are researching your animal, use this checklist to make notes about its mental ability.

Sign of Intelligence	Is it Present?	Notes or Examples	Source
Memory	Yes or No		
Problem Solving	Yes or No		
Tool Use	Yes or No		
Math	Yes or No		
Self-aware	Yes or No		
Dreaming or Imagination	Yes or No		
Sense of Time	Yes or No		
Instinct	Yes or No		
Personality	Yes or No		
Empathy	Yes or No		

LESSON 17

Zooming in on Zoos

Objectives

▶ Students will research two or three zoos that have the animal they are researching in captivity.

▶ Students will create a sketch/rough draft of what they will create for their model.

Materials

▶ *Zooming in on Zoos* handout

▶ *Need to Know Board* handout

▶ *A 20 Question Creature Quest* handout

 DOI: 10.4324/9781003334743-18

Assessments

▶ *Zooming in on Zoos* handout
▶ Journal prompt

Procedures

1. Review major concepts from the last class period (e.g., instinct versus learned behavior, other signs of animal intelligence, etc.).

2. Remind students of the differences between modern zoos and those of the past.

3. Explain to students that their task today is to locate two or three zoos around the world that have the animal they are researching on display.

4. Distribute the *Zooming in on Zoos* handout and go over the directions.

5. Provide students with the website listed below. It lists 50 of the best zoos in the world along with a link to each zoo's website. Remind students to research their local zoo as well.

   https://tourscanner.com/blog/best-zoos-in-the-world/.

6. Monitor and support students as they work. Encourage students to help each other by pointing out if a zoo they are researching has a great exhibit of an animal their classmate is researching.

7. Debrief with students regarding how their planning is going and what type of final product they will create.

8. Go over the project checklist included below:

   a. The zoo exhibit must include living and nonliving things interacting together.

   b. The zoo exhibit should resemble the animal's natural habitat.

   c. The zoo exhibit should be large enough for the ecosystem to thrive.

   d. The zoo exhibit should be innovative and creative.

   e. The zoo exhibit must be realistic and possible to build within an actual zoo.

9. Journal prompt: "If you could visit any zoo in the world, which one would you choose and why?".

Zooming in on Zoos

Name _____   Date _____

PART 1

Directions: Find two or three zoos around the world that have the animal you have been researching in captivity. Record the location and name of each zoo. Next, locate a picture of the actual exhibit or read the description on the zoo's website. Take notes on anything you would like to include in your exhibit. Remember, your task is to recreate your animal's ecosystem in a natural way. Your exhibit should mimic your animal's habitat in the wild.

	Location of the Zoo	*Name of the Zoo*	*Positive Elements You Might Use*
Zoo 1			
Zoo 2			
Zoo 3			

PART 2

Directions: Consult *A 20 Question Creature Quest handout* and pay special attention to questions 6, 7, and 8. Use the back of this paper to sketch out a possible design for your zoo exhibit. Make sure to label each living and nonliving thing you plan to include.

PART 3

Directions: Decide if your zoo exhibit will be a physical diorama, a digital diagram, or some other type of product. Once this has been approved by your teacher, create a list of materials you might need to build your model.

LESSON 18

Creating Communities, Not Cages

Objectives

▶ Students will continue planning and begin construction of their animal ecosystem zoo exhibit.

▶ Students will be introduced to the final piece of the vertebrate animal problem-based learning task (i.e., creating an enrichment toy for their animal).

Materials

▶ Individualized list of materials based on student projects

▶ Completed *Zooming in on Zoos* handout

▶ *Need to Know Board* handout

▶ *A 20 Question Creature Quest* handout

DOI: 10.4324/9781003334743-19

Assessments

▶ Beginning stages of their exhibits
▶ Journal prompt

Procedures

1. Review major concepts from the last class period (e.g., best zoos around the world, materials they will need to create their exhibit, etc.).

2. Explain to students that their main task today is to finalize their zoo exhibit plans and either begin construction of the final product or continue gathering materials.

3. Emphasize again the differences between communities of animals living in ecosystems versus the old-fashioned cages found in outdated zoos.

4. Give students around 20 minutes to work independently on their projects.

5. Pause the students and ask them to find a partner, share their ideas or drawings, and give each other feedback on their projects.

TEACHER'S NOTE

It is important that students know how to give proper feedback to each other. Consider a mini-lesson that includes discussing how each of these two statements make them feel:

• This is just confusing and sounds dumb. That drawing of yours is ugly, too.
• I am not sure what you mean by _____ Maybe we could find a picture online you could copy and paste into your document together.

6. Give students several more minutes to work independently to incorporate their partner's suggestions, ideas, etc.

7. Ten minutes before class ends, pull students together and present the final phase of the vertebrate animal problem-based learning task using the script below:

> In addition to designing a new exhibit for your animal, we would like you to design an enrichment toy it might enjoy playing with. An example of an enrichment toy is a Kong for dogs. This is a hollow shape made of rubber that the owner fills with peanut butter and treats. The dog has to find a way to get the treats out of the toy (show the pictures). A second example is a laser pointer for a cat.

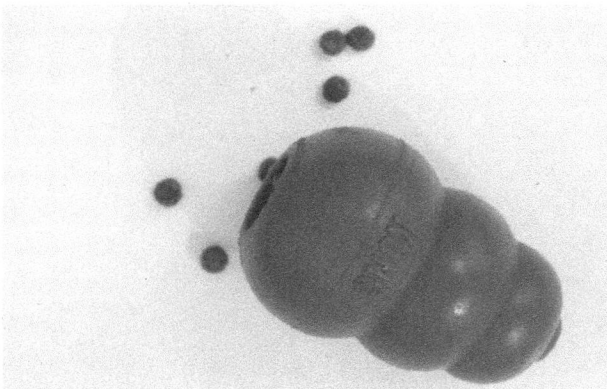 

8. Work together as a class to create an idea for a rabbit as practice. The first step is to find out what rabbits like to do, eat, etc. Ask the students to share facts they know about rabbits (e.g. they like to eat carrots, they hop, etc.). The second step is to brainstorm possible ideas. Remind students of the rules for brainstorming (i.e., all ideas count, piggy-backing ideas is encouraged, the more ideas the better) and then list ideas on the board together.

9. Decide on the best idea as a group and sketch out what it might look like.

10. Ask students to revisit their *Need to Know Boards* and add information based on this new requirement.

11. Tell students their homework is to bring in any materials they need for their projects and begin thinking about a possible toy their animal would enjoy.

12. Journal prompt: "Why is it good for animals to play with each other or with toys?".

Image Credits

Bokehboo Studios/Shutterstock.com
nataliajakubcova/Shutterstock.com

Kids at Work

Objectives

▶ Students will finalize construction of their animal ecosystem zoo exhibits and plans for an innovative enrichment toy.

▶ Students will assess their products using two checklists and prepare for a presentation.

Materials

▶ Individualized list of materials based on student projects

▶ Checklist included in step 3

DOI: 10.4324/9781003334743-20

Assessments

▶ Final zoo exhibit projects

▶ Journal prompt

Procedures

1. Review major concepts from the last class period (e.g., what an enrichment toy is for, why play is important, etc.).

2. Explain to students that their main task today is to finalize the construction of their zoo exhibit product and enrichment toy idea.

3. Remind students of the following criteria for their zoo exhibits:
   a. The zoo exhibit includes living and nonliving things interacting together.
   b. The zoo exhibit resembles the animal's natural habitat.
   c. The zoo exhibit would be large enough for the ecosystem to thrive.
   d. The zoo exhibit is innovative and creative.
   e. The zoo exhibit is realistic and possible to build within an actual zoo.

4. Tell students that they will also need to draw a labeled picture of their enrichment toy and write a paragraph describing it. Their toy should:
   a. Provide a treat your animal will enjoy.
   b. Be safe for the animal.
   c. Be strong and not easily destroyed by the animal.

5. Provide time for students to work. Monitor, question, and support as this takes place.

6. A few minutes before class is over, explain to students that they will present their final products to the class during the next class period. Any final work needs to be completed by then.

7. Journal prompt: "I feel _____ about my project because _____".

Presentations Aplenty

Objectives

▶ Students will present their final zoo exhibits and toys to the class.

▶ Students will give feedback to their peers and reflect on their own presentations.

Materials

▶ Final zoo exhibits

▶ Diagram of labeled enrichment toy with descriptive paragraph

 DOI: 10.4324/9781003334743-21

Assessments

▶ Final zoo exhibit projects

▶ Diagram of labeled enrichment toy with descriptive paragraph

▶ Journal prompt

Procedures

1. Review major concepts from the last class period (e.g., major requirements of the projects, etc.).

2. Explain to students that today they will present their project to the class using the following sentence frames:

   a. My animal was …

   b. I chose this animal because …

   c. Three things I learned about my animal were …

   d. My zoo exhibit is innovative because …

   e. This is an ecosystem because I included …

   f. My enrichment toy would be fun for my animal because …

3. Give students five minutes to prepare how they will answer the sentence frames.

4. List the characteristics of a good presenter and remind students to exhibit them as they present:

   a. Speak so everyone can hear you.

   b. Make eye contact with the audience.

   c. Don't bury your face behind a piece of paper or book.

   d. Answer questions at the end.

TEACHER'S NOTE

Many students are visual learners. Consider showing students a short video clip of someone who is an excellent presenter followed by a clip of someone who is not. Ask students to compare the two clips and point out what the major differences were.

5. Ask for a volunteer and begin the presentations. Pause for questions between presenters.

6. Once everyone has presented, ask the students to find a partner and discuss the two questions below:

    a. My favorite part about your project was ...

    b. One thing you might think about adding or changing is ...

7. Instruct students to leave their exhibits at school for one more day.

8. Journal prompt: "I think my presentation was _____. I think this because _____."

Structuring the Spineless

Objectives

▶ Students will collaboratively arrange their individual zoo exhibits to create one class zoo.

▶ Students will construct a dichotomous key useful for identifying invertebrate animals.

Materials

▶ Final zoo exhibits from each student

▶ *The Invertebrates' Plea!* poem from Lesson 3

▶ *The Five Kingdoms* chart from Lesson 3

▶ One piece of posterboard

 DOI: 10.4324/9781003334743-22

▶ Marker
▶ Print out of the *Invertebrate Dichotomous Key Cards*

Assessments

▶ Invertebrate Dichotomous Key performance task
▶ Journal prompt

Procedures

1. Review major concepts from the last class period (e.g., presentation reflections, etc.).

2. Section off an area of the floor or a large table to serve as the land on which the students will construct a class zoo.

3. Ask the students to talk with one another to decide the best way to organize the zoo exhibits. For example, most zoos have a section for reptiles, a section for African mammals, a section for birds, etc. Show students a map of their local zoo either online or in print.

4. Challenge the students to work together to place the zoo exhibits in the most appropriate but innovative way they can. Give students 15 minutes to physically place the exhibits and then justify their reasoning with each other.

5. Once everyone has come to a consensus, ask the students to generate a fun name for our newly created zoo.

6. Shift gears by explaining to students that they will now spend several days talking about invertebrate animals. Ask students if anyone remembers what an invertebrate is from Lesson 3.

7. Reread the poem *The Invertebrates' Plea!* and show *The Five Kingdoms* chart from Lesson 3.

8. Explain to students that we are now going to see if we can create a dichotomous key together that will reveal to us the structure of invertebrate animal groups.

9. Recreate the blank template shown here either on the white board, screen, or on a large piece of posterboard. (NOTE: Each of the rectangles should be the same size as the vertebrate dichotomous key cards.)

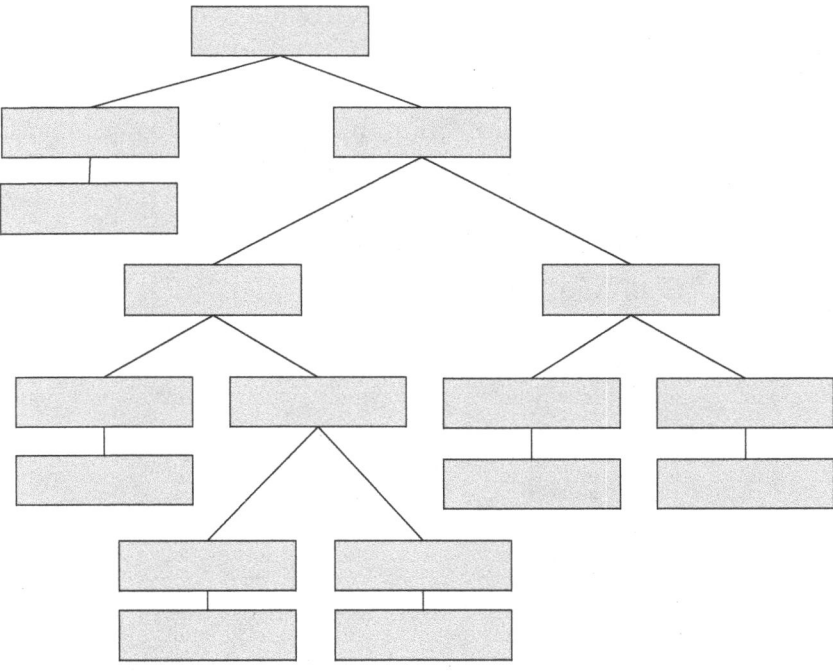

10. Print and cut out the premade dichotomous key cards included below. (NOTE: When printing, it may help your students complete this task if the descriptors are printed on one color paper and the names of the animal groups are printed on another.)

11. Ask students if they know what the terms below mean. If not, discuss the definitions provided.

    a. *symmetrical* – parts that are the same in size and shape

    b. *bilateral symmetry* – when only one line can divide a shape into identical halves

    c. *radial symmetry* – multiple lines can divide a shape into similar halves (like rays of the sun or pieces of a pie)

    d. *segmented* – repeating sections joined together to form something

    e. *exoskeleton* – a skeleton (like armor) on the outside of an animal

12. Challenge students to work together to place the invertebrate dichotomous key cards on the template in a way that makes sense. Remind students of the *guess and check* strategy, which means they may need to rearrange or start over several times until they find a solution that makes sense. The correct answers are shown on the next page.

13. Once students have figured out the correct structure, name a common invertebrate animal and ask them to use the key to sort it into the correct group.

14. Challenge each student to name an invertebrate animal of their own and verify the accuracy of the key by seeing if it works to sort the animal they chose.

15. Journal prompt: "Which invertebrate animal group are you the most curious about? Why?".

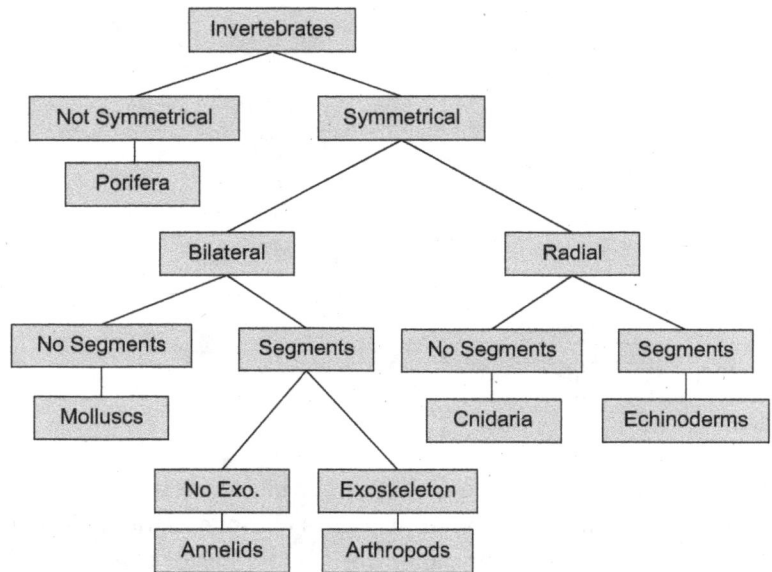

TEACHER'S NOTE

Consider inviting other classes or parents/community members to visit your classroom to see your class zoo. This would be an excellent opportunity for students to present in front of an authentic audience, hone their speaking skills, and educate others.

Invertebrate Animal Dichotomous Key Cards

Invertebrates
Porifera
Cnidaria
Arthropods
Annelids
Echinoderms
Molluscs
Symmetrical
Symmetrical
Bilateral
Radial
Segmentation
Segmentation
Segmentation
Segmentation
Exoskeleton
Exoskeleton

Investigating Invertebrates

Objectives

▶ Students will independently explore the world of invertebrates.

▶ Students will complete a project-based learning task from a menu.

Materials

▶ PDF of *The H.O.T. Spot: Invertebrates*: https://drive.google.com/file/d/1igAtgR FFe4vhLdowAs4x_ghyigqWgRh0/view?usp=sharing

▶ Headphones for each student

DOI: 10.4324/9781003334743-23

Assessments

▶ Project-based learning task

▶ Journal prompt

Procedures

1. Review major concepts from the last class period (e.g., the components of a dichotomous key for invertebrate animals.).

2. Explain to students that they will have the opportunity to learn more about the invertebrate animal group independently today.

3. Provide a digital copy of the linked *H.O.T. Spot* to each student. Headphones are also recommended.

4. Explain to students that they will have the entire class period to explore the content included on the *H.O.T. Spot*. The MUST do sections to complete are:

   a. The *Did You Know* section

   b. The *Help Wanted* section

   c. One of the three *Project-Based Bonanza* tasks

   d. One of the two writing prompts

5. Monitor students as they work, provide feedback, and ask probing questions.

6. Ten minutes before the end of the class, pull students back together and answer the *Digging Into Depth and Complexity* questions/prompts as a class.

7. Provide an opportunity for students to share their completed project-based learning task from page three.

8. Journal prompt: "Tomorrow, we will begin a fun research project in which you will have to choose four invertebrates to explore. Which invertebrates might you consider adding to your list?".

TEACHER'S NOTE

Project-based learning and problem-based learning are not the same thing. Project-based learning involves presenting a defined task to students they must then complete. Problem-based learning involves presenting a messy problem to students that they must then analyze, determine next steps, and solve using an inquiry approach.

No Spine Battle Lines

Objectives

▶ Students will be introduced to a new problem-based learning task.

▶ Students will select four invertebrates and begin researching their salient characteristics.

Materials

▶ *No Spine Battle Lines* sample cards

▶ Blank card templates for students

▶ *Need to Know Board* handout (see Lesson 7)

 DOI: 10.4324/9781003334743-24

Assessments

▶ *Need to Know Board* handout

▶ Journal prompt

Procedures

1. Review major concepts from the last class period (e.g., various facts learned from yesterday's H.O.T. Spot, etc.).

2. Ask the students if anyone has ever played with or collected trading cards (e.g., Pokémon, baseball, etc.).

3. Introduce a new problem-based learning task using the script below:

   Imagine the makers of Pokémon decided to create a new game that required the collection of invertebrate cards instead of pretend cartoon creatures. You just bought the starter set (*pass out the six example invertebrate cards included*). The game is played just like Pokémon. Each person plays one card at the same time and the points determine whose card wins and whose card loses (show diagram below).

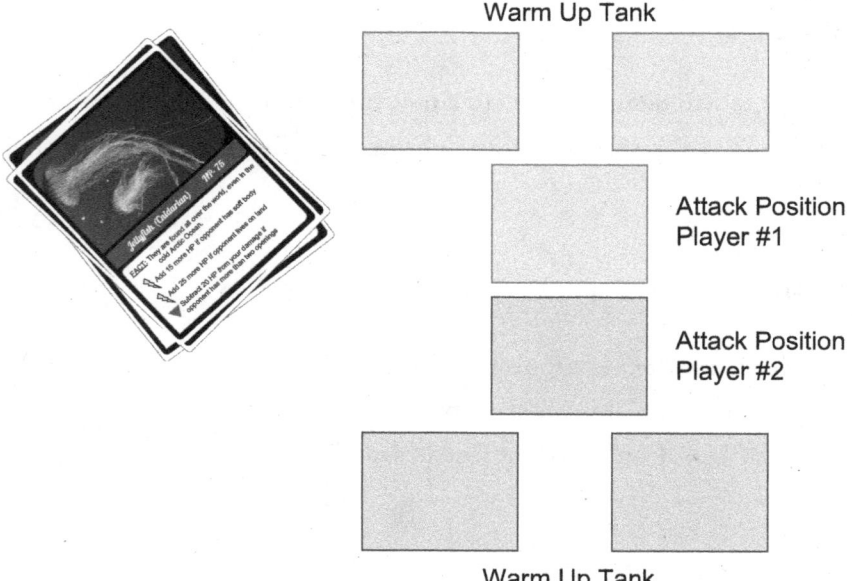

Your task over the next two class periods is to create at least four more cards to add to the set. The company will decide if they will be included in future versions of the game. If they are quality cards, you will be rewarded for your work. If not, you will be asked to make revisions.

Remember, all invertebrates can be classified into one of six groups. Also, when making your cards, the characteristics of each invertebrate should determine its starting HP (Hit Points), which other invertebrates it can successfully attack, what its weaknesses are, etc.

4. Give each student a fresh *Need to Know Board* and ask them to record what they *know* and what they *need to know* in order to successfully complete this task.

5. Provide the rest of class for students to begin choosing the invertebrates they would like to use and/or researching important facts about their chosen invertebrates.

6. Monitor and support students as they work. When possible, ask students to choose a specific species of an animal instead of a broad type like the sample cards.

7. A few minutes before class ends, pull students back together and debrief their progress.

8. Journal prompt: "You will have the entire next class period to continue working on this project. What will be your plan of action when we return to class once more?".

Sample No Spine Battle Line Cards

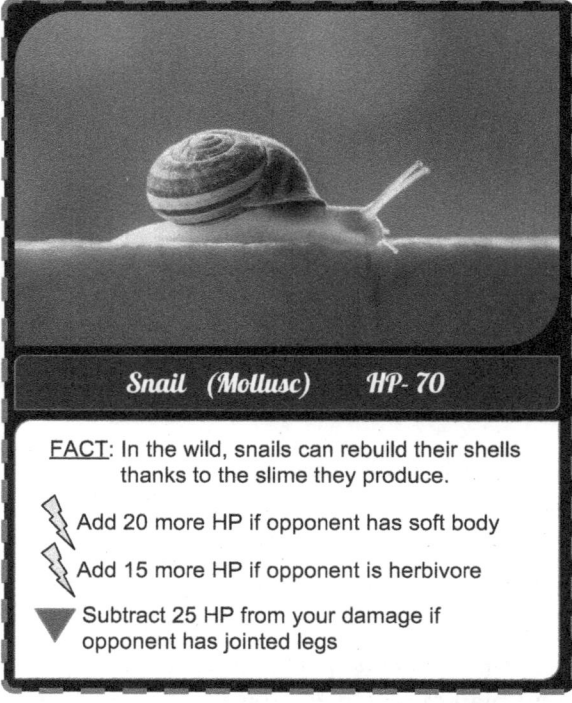

Snail (Mollusc)   HP- 70

FACT: In the wild, snails can rebuild their shells thanks to the slime they produce.

⚡ Add 20 more HP if opponent has soft body

⚡ Add 15 more HP if opponent is herbivore

▼ Subtract 25 HP from your damage if opponent has jointed legs

Arachnid (Arthropod)   HP- 80

FACT: All spiders have venom, but only a few are dangerous to humans.

⚡ Add 10 more HP if opponent is arthropod

⚡ Add 15 more HP if opponent is annelid

▼ Subtract 35 HP from your damage if opponent lives in the sea

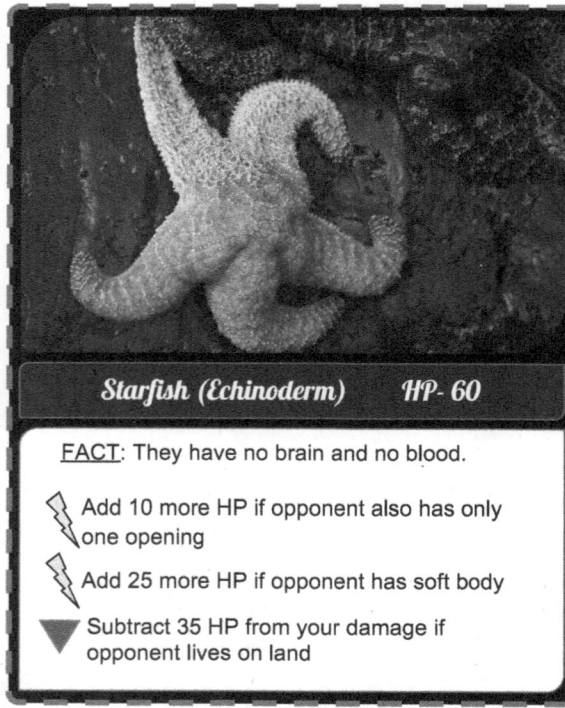

Starfish (Echinoderm)   HP- 60

FACT: They have no brain and no blood.

⚡ Add 10 more HP if opponent also has only one opening

⚡ Add 25 more HP if opponent has soft body

▼ Subtract 35 HP from your damage if opponent lives on land

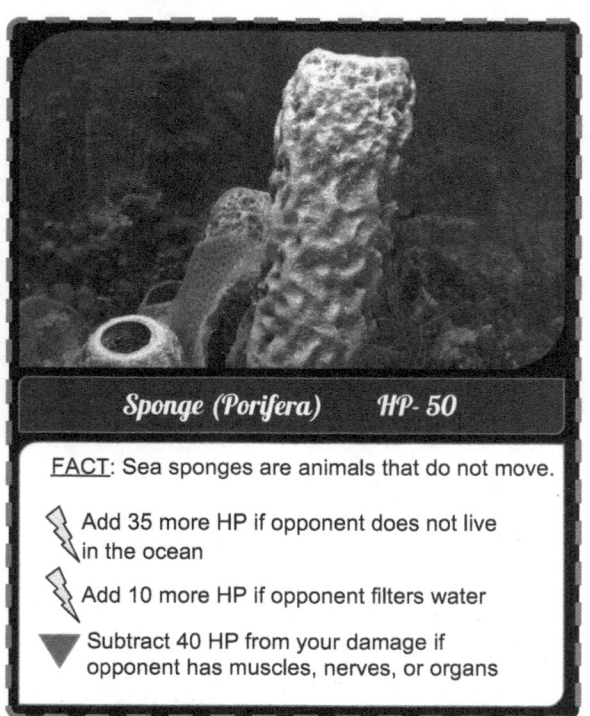

Sponge (Porifera)   HP- 50

FACT: Sea sponges are animals that do not move.

⚡ Add 35 more HP if opponent does not live in the ocean

⚡ Add 10 more HP if opponent filters water

▼ Subtract 40 HP from your damage if opponent has muscles, nerves, or organs

LESSON 24

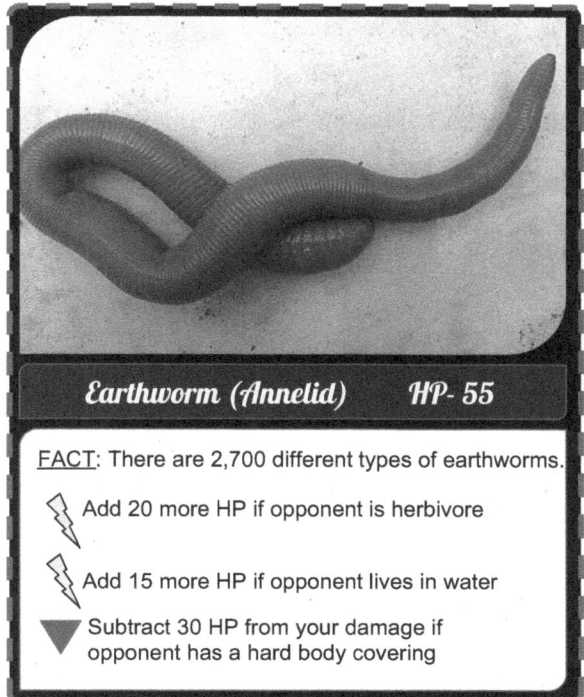

Earthworm (Annelid)   HP- 55

<u>FACT</u>: There are 2,700 different types of earthworms.

⚡ Add 20 more HP if opponent is herbivore

⚡ Add 15 more HP if opponent lives in water

▼ Subtract 30 HP from your damage if opponent has a hard body covering

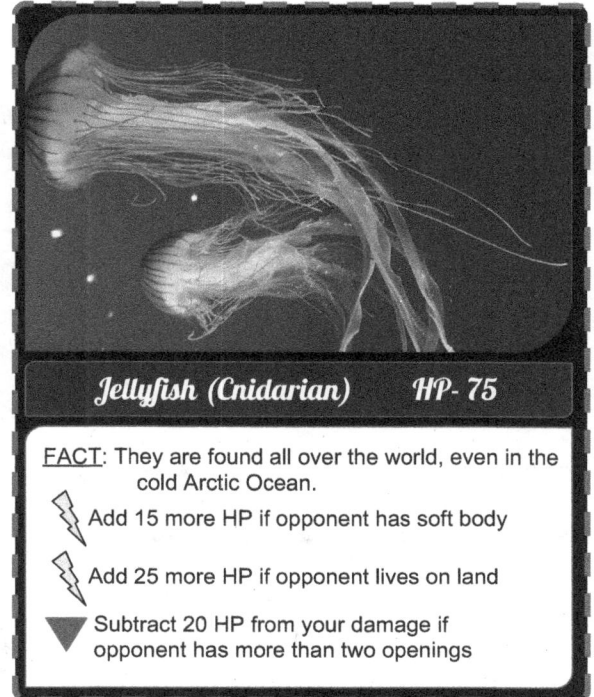

Jellyfish (Cnidarian)   HP- 75

<u>FACT</u>: They are found all over the world, even in the cold Arctic Ocean.

⚡ Add 15 more HP if opponent has soft body

⚡ Add 25 more HP if opponent lives on land

▼ Subtract 20 HP from your damage if opponent has more than two openings

Card Template for Students

Card 1

Creature's Name	Type of Invertebrate
_____ (	) HP- ___

FACT:

⚡ Add ___ more HP if…

⚡ Add ___ more HP if…

▼ Subtract ___ HP from your damage if…

Card 2

Creature's Name	Type of Invertebrate
_____ (	) HP- ___

FACT:

⚡ Add ___ more HP if…

⚡ Add ___ more HP if…

▼ Subtract ___ HP from your damage if…

Card 3

Creature's Name	Type of Invertebrate
_____ (	) HP- ___

FACT:

⚡ Add ___ more HP if…

⚡ Add ___ more HP if…

▼ Subtract ___ HP from your damage if…

Card 4

Creature's Name	Type of Invertebrate
_____ (	) HP- ___

FACT:

⚡ Add ___ more HP if…

⚡ Add ___ more HP if…

▼ Subtract ___ HP from your damage if…

Image Credits

Snail photo © Waugsberg / Wikimedia Commons / CC-BY-SA-3.0
Spider photo by David Mark from Pixabay
Starfish photo © Steven Pavlov / http://commons.wikimedia.org/wiki/User:Senapa
Sponge photo by Timothy W. Brown, Public domain, via Wikimedia Commons
Earthworm photo by Alex Popovkin, Bahia, Brazil from Brazil, CC BY 2.0, via Wikimedia Commons

The Work Continues

Objectives

▶ Students will complete four new *No Spine Battle Lines* cards.

▶ Students will play the game with a partner.

Materials

▶ *No Spine Battle Lines* cards

▶ *Need to Know Board* handout (see Lesson 7)

 DOI: 10.4324/9781003334743-25

Assessments

▶ Completed *No Spine Battle Lines cards*

▶ Journal prompt

Procedures

1. Review major concepts from the last class period (e.g., the instructions for their invertebrate problem-based learning task, etc.).

2. Ask the students how they are feeling about their invertebrate cards.

3. Explain to students that it is very important to have realistic Hit Point values assigned to each animal. Give the following examples:

    a. A slug should have a low Hit Point value because it is not a predator, moves slowly, and has few defenses.

    b. A hornet could have a higher Hit Point value because it can sting, flies very quickly, and can be aggressive.

4. Ask a few students to share the names of their invertebrate animals and the initial Hit Points they have assigned to them.

5. Guide students in adjusting their totals to reflect reality. In general, the Hit Points should range from 25 to 150. Only one of their invertebrate animals should have Hit Points greater than 100.

6. Release students to work on completing their cards.

7. Pull students back together 15 minutes before the end of class.

8. Tell students we will use similar rules to Pokémon, except we will use only three cards set up like the diagram from yesterday. Ask students to read how to play Pokémon using the link here: www.wikihow.com/Play-With-Pok%C3%A9mon-Cards.

9. Provide time for students to play one round of *No Spine Battle Lines*. Explain to students that they will have time to play additional rounds during the next class period.

10. Journal prompt: "Do you think *No Spine Battle Lines* would be a bestseller in stores? How are some invertebrates similar to Pokémon creatures?".

The Invincible Invertebrate

Objectives

▶ Students will create a hybrid invertebrate animal combining features from three or more invertebrate groups.

▶ Students will participate in the Tournament of Champions using their newly created creatures.

Materials

▶ Blank *No Spine Battle Lines* card template

▶ Completed *No Spine Battle Lines* cards

▶ *Need to Know Board* handout (see Lesson 7)

DOI: 10.4324/9781003334743-26

INVALUABLE INVERTEBRATES and Species with Spines

Assessments

▶ Fifth completed card
▶ Journal prompt

Procedures

1. Review major concepts from the last class period (e.g., how to play the game, etc.).

2. Explain to students that today will be broken up into two parts. At the beginning of class, they will create one more special invertebrate card that will be called *The Invincible Invertebrate*. During the second half of class students' invincible invertebrates will battle each other in a tournament style format.

3. Provide the following guidelines for their invincible invertebrate card:

   a. This card should feature a pretend creature that you create in your secret laboratory.

   b. It should be a hybrid (or combination) of characteristics from three or more invertebrate groups (e.g., arthropod + annelid + cnidarian).

   c. The Hit Point value should range from 200 to 300.

   d. The name of the creature, interesting fact, and picture should be invented by you.

   e. Students should keep their creations top secret from their classmates until the appointed time.

4. Once students have finished creating their Invincible Invertebrate cards, announce the *Tournament of Champions*. This will consist of each student going into a head to head battle with every other student in the classroom one at a time. A "battle" means showing each other your card, comparing the two, and determining which creature would have the highest number of points after deductions and additions are made. The process will repeat again with the next classmate. Students should keep track of how many times their Invincible Invertebrate card wins using tally marks in their journals. Each round should take only two or three minutes.

5. Check to make sure every student had the opportunity to battle each other. Next, ask each student to share the total number of times their creature beat their opponents. The person who has the highest number of wins is the ultimate invincible invertebrate. In case of a tie, have the top scorers compare their cards. The winner of that round takes first place.

6. Ask students to reflect on this process and discuss the following question: *What external or internal structures were most useful to surviving an attack from an opponent?*

7. Allow students the rest of class to play the full version of the game using all of their cards.

8. Journal prompt: "Sometimes we win and sometimes we lose. It is important to be humble when we win and to not be devastated if we lose. How do you typically deal with winning and losing?".

TEACHER'S NOTE

You will need to collect all invincible invertebrate cards and duplicate them for the next lesson. You should create four sets of each.

Animal I.D. Key

Objectives

▶ Students will identify unknown animals using a dichotomous key.

▶ Students will create their own dichotomous keys for the *Invincible Invertebrate* cards they created during the last class period.

Materials

▶ The blank vertebrate dichotomous key with cards (see Lesson 5)

▶ The blank invertebrate dichotomous key with cards (see Lesson 22)

▶ *Animal I.D. Key* handout

▶ Animal picture cards

 DOI: 10.4324/9781003334743-27

- ▶ Blank paper or posterboard
- ▶ The *Invincible Invertebrate* cards from last class period

Assessments

- ▶ *Animal I.D. Key* handout
- ▶ Student created *Invincible Invertebrate* dichotomous keys
- ▶ Journal prompt

Procedures

1. Review major concepts from the last class period (e.g., *Tournament of Champions, Invincible Invertebrate* cards, etc.).

2. Explain to students that today we will revisit the idea of dichotomous keys. Divide the class into two groups. Give one group the vertebrate dichotomous key with cards and the other the invertebrate dichotomous key. Challenge them to place the cards in the proper boxes just as we did in prior lessons.

3. Next, ask the students to step away from their work and swap places. Their task is to check the work of the other group and make any necessary changes to the structure.

4. Carefully place the two keys right next to each other. This should make one large dichotomous key if you add one more heading. Show students by adding the heading *Kingdom Animalia*.

5. Give students a copy of the *Animal I.D. Key* handout and the animal picture cards.

6. Ask students to work in small groups to identify the names of the mystery animals using the key provided.

7. Go over the correct answers by asking students to share their conclusions.

8. Ask students to discuss the following question with their groups: *Which were the easiest to identify? Which were the most difficult? Why?*

9. Share with students that you recognize the task you gave them was not very challenging. This is because they had a limited number of animal cards and a very specific key.

10. Explain to students that their task for the rest of the class period is to create their own dichotomous keys for the *Invincible Invertebrate* cards they created during the last class. Give each small group of students a piece of posterboard or drawing paper and a set of the cards.

11. List the characteristics of a quality dichotomous key:

    a. Each branch of the key clearly subdivides the cards into smaller groups.

    b. The end point of each branch of the key leads to ONE and only one card.

    c. The cards are relatively easy to subdivide into groups using the criteria on the key and the creature cards provided (e.g., has an exoskeleton, does NOT have an exoskeleton, etc.).

12. A few minutes before the class period is over, ask students to share the organizing structure they used to create their keys. Explain to students that they will have time to finish this task during the next class period.

13. Journal prompt: "Do you prefer making a dichotomous key or using a dichotomous key? Why?".

Animal I.D. Key

Name _____   Date _____

Directions: Please use the dichotomous key below to identify the unknown animal cards given to you by your teacher.

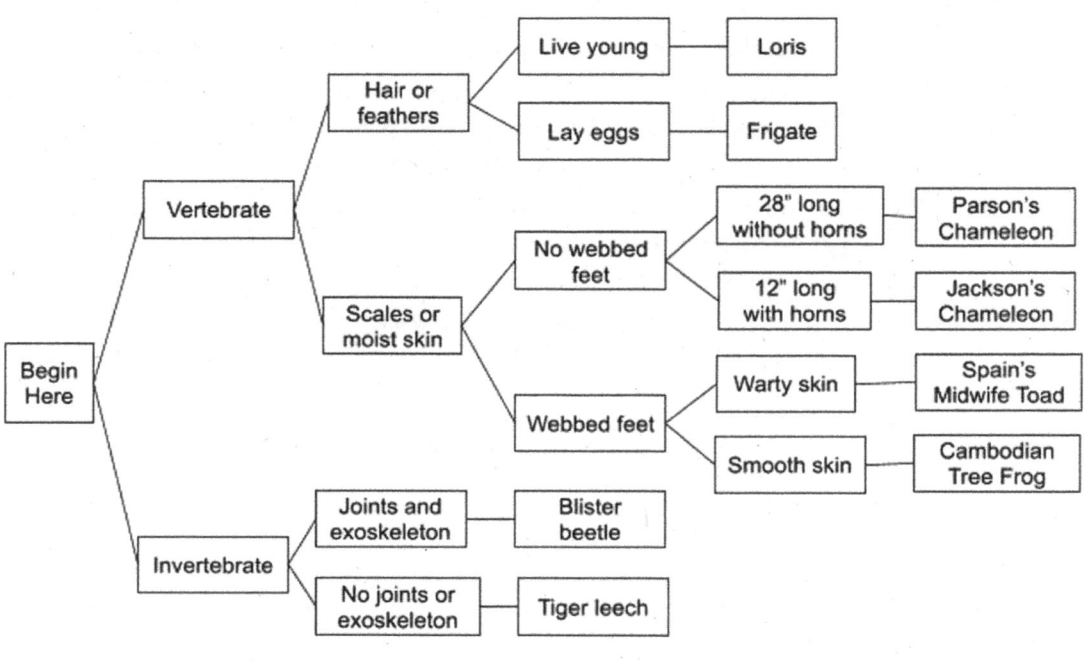

Creature 1 = _____

Creature 2 = _____

Creature 3 = _____

Creature 4 = _____

Creature 5 = _____

Creature 6 = _____

Creature 7 = _____

Creature 8 = _____

LESSON 27

Animal Cards

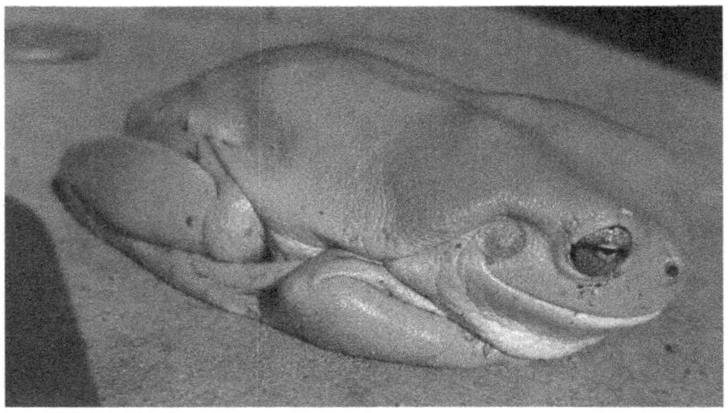

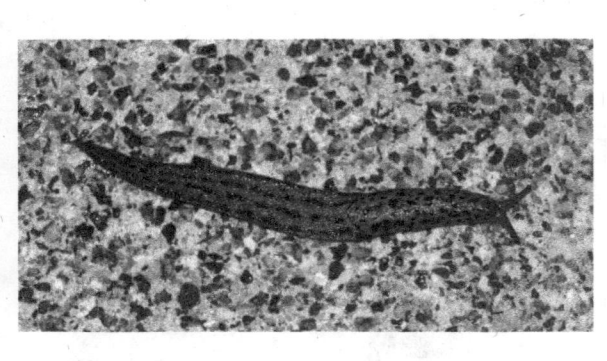

Image Credits

Alex Pyron, CC BY-SA 3.0, via Wikimedia Commons

vagabond54/Shutterstock.com

Critiquing Your Creature Keys

Objectives

▶ Students will evaluate the accuracy and effectiveness of the dichotomous keys they created in the last class period.

▶ Students will choose a task from a choice menu to complete based on interest.

Materials

▶ Student-created dichotomous keys from last class period

▶ The *Invincible Invertebrate* cards from last class period

▶ 3 × 5 card for each student

▶ Choice menu

DOI: 10.4324/9781003334743-28

Assessments

▶ Peer evaluations of dichotomous keys

▶ Completed choice menu tasks

▶ Journal prompt

Procedures

1. Review major concepts from the last class period (e.g., criteria for making a dichotomous key, etc.).

2. Give students a few minutes to finalize the creation of their dichotomous keys from yesterday.

3. Ask students to take out the copies of the creature cards you gave them while creating the dichotomous keys. Instruct students to use a black marker to cover up the name and of each creature and the type of invertebrate. (See diagram below.)

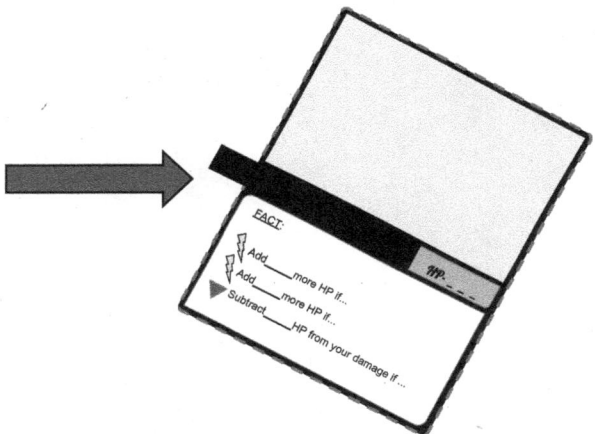

4. Students should give their set of cards and the dichotomous keys they created to another group. Challenge students to identify the names of each creature using the dichotomous keys of their classmates.

5. When finished, give each student a 3 × 5 card. Ask students to write constructive feedback about their classmates' keys using the criteria provided in the last class period. Which criteria did they meet? Which criteria did they not?

6. Each group should return the creature cards, dichotomous keys, and 3 × 5 cards to the original owners. Ask the original creators to read their feedback and make any necessary changes to improve the final result.

TEACHER'S NOTE

This is the perfect opportunity to introduce the idea of *iterations*. An iteration is when a product improves when each new update or version is released. A perfect example is Apple's iPhone. The latest iPhone looks nothing like the first version. Explain to students how they can apply this concept to their own work.

7. When students are finished, distribute the choice menu below. Their task today and during the next class period is to complete the required task plus one additional square.

8. Journal prompt: "How did it feel to evaluate another group's work? How did it feel to receive feedback from your peers?".

Animals, Habitats, and Ecosystems

Choice Menu

Name _____   Date _____

Directions: Please complete the Required Task and at least one additional task.

REQUIRED TASK	CHOICE A	CHOICE B
 On day one of this unit, you completed a pretest. The last question on the test was "What question do you have about animals?". Your teacher will tell you what you wrote down. Research the answer to your question. If your question has already been answered, list something else you would like to know and find the answer.	Find out more about a few of the zoologists/biologists listed below: • Jane Goodall • Con Slobodchikoff • Fredrick Frohawk • Ayana E. Johnson • David Attenborough • Georges Cuvier • Mary Rathbun What did they research? How did they make a difference? What challenges did they face?	Read a book with an environmental theme, such as: • *Hoot* by Carl Hiaasen • *The Plastic Problem* by Aubre Andrus • *Heroes of the Environment* by Harriet Rohmer • *A Wolf Called Wander* by Rosanne Parry Was the book fiction or nonfiction? What did you learn about human impacts on the environment? What will you do differently as a result?
CHOICE C	**CHOICE D**	**CHOICE E**
To migrate means to relocate from one habitat to another on a regular cycle. Many species of animals migrate each year, but birds do it spectacularly well and in large numbers. In fact, 4,000 species (40% of all birds) are migratory. Research five birds that migrate and then draw their routes on a map.	The insect family is so successful that 90% of all Earth's creatures belong to this group. They are so plentiful that there are more than 9,000 different types of ants! Research five or more species of ants and then create a small dichotomous key that would be useful for telling them apart. Next, draw ant picture cards listing facts that a friend could use to sort into species.	How much do you know about ocean life? Take this quiz to find out. Oh, by the way, you will NOT find an answer key. You will have to do the research yourself to see if you are correct! 1. How many lightbulbs could an electrical eel power with one sting? 2. Which have been around longer: jellyfish or dinosaurs? 3. What color is the blood of an octopus? 4. What sea creature has no head, mouth, eyes, feelers, bones, heart, lungs, or brain, but is still alive?

Image Credit
Frank Heikkinen/Shutterstock.com

LESSON 28

Let's Review Before We're Through

Objectives

▶ Students will review content covered during the unit.

▶ Students will choose a task from the choice menu to complete based on interest.

Materials

▶ Choice menu

 DOI: 10.4324/9781003334743-29

Assessments

- ▶ Review games
- ▶ Completed choice menu tasks
- ▶ Journal prompt

Procedures

1. Review major concepts from the last class period (e.g., answers to questions they have about animals, etc.).

2. Explain to students that today they will review for the post-test they will take during the next class period and then continue to work on the choice menu tasks from the previous lesson.

3. Ask students to discuss the following prompt with a partner: *The most important thing I learned about animals during this unit was ...*

4. Give students the opportunity to share their responses with the entire class.

5. Play one of the review games as a class listed below:

   a. **Mystery Animal**: Ask the students to secretly choose an animal they know a lot about without telling anyone what it is. Ask one student to volunteer to come to the front of the room. The class must ask this person *yes or no* questions about the traits of the volunteer's animal using terms we have learned during this unit (e.g., diurnal, nocturnal, omnivore, carnivore, etc.) until someone can guess its identity correctly. Continue until all students have gone.

   b. **Tic Tac Know**: Divide students into two teams. One team will be the X's and one team will be the O's. Draw a Tic Tac Toe board on the white board at the front of the classroom. Pose a question to each team, one at a time. If the team gets the answer correct, they get to add an X or O to the board. Continue until a winner is declared.

6. Give students an opportunity to ask any questions they still have about content covered during the unit.

7. Provide the rest of the class period for students to continue working on their choice menu from the previous class period.

8. Journal prompt: "My favorite part of the unit ... "

Show What You Know

Objectives

▶ Students will reflect on the unit.
▶ Students will take the posttest for the unit.

Materials

▶ Completed *Need to Know Boards* from the unit
▶ Student journals
▶ Post-test (same assessment as Lesson 1)
▶ Choice menu

   DOI: 10.4324/9781003334743-30

Assessments

▶ Post-test

Procedures

1. Ask the students what it means to *reflect*. Define this as *thinking deeply or carefully about something*.

2. Instruct students to look through their *Need to Know Boards* and journal responses completed during the unit. Challenge students to answer the following three questions as a class:

   a. Why was it important for you to learn about these concepts?

   b. What do you still want to learn about animals?

   c. How could my teacher make the unit better in the future?

3. Remind students of the pretest on day one. Explain that today they will take the same test in order to measure how much they have learned.

4. Distribute the post-test and ask students to complete it to the best of their ability.

5. When everyone has finished, collect the post-tests and go over the correct answers together.

6. Next, distribute the pretests from the beginning of the unit and ask students to compare their performance from pre to post.

7. If time permits, ask students to finalize their choice menu projects or begin a new one.

References

Jenkins, M. (2001). *Chameleons are cool*. Candlewick.

Jenkins, S. (2001) *What do you do when something wants to eat you*. Clarion Books.

Madden, D. (1993). *Wartville wizard*. Aladdin.

National Association for Gifted Children. (2019). 2019 Pre-K–Grade 12 gifted programming standards. www.nagc.org/sites/default/files/standards/Intro%202019%20Programming%20Standards.pdf

National Governors Association Center for Best Practices, & Council of Chief State School Officers. (2010). *Common core state standards*. Authors.

NGSS Lead States. (2013). *Next generation science standards: For states, by states*. The National Academies Press.

Prelutsky, J. (1983). *Zoo doings*. Scholastic Press.

Schwartz, D. (1999). *If you hopped like a frog*. Scholastic Press.

Master Materials List

Lesson 1

▶ Notebook for each student

Lesson 5

▶ One piece of posterboard

Lesson 6

▶ Dice

Lesson 7

- ▶ Nonfiction books about animals

Lesson 9

- ▶ Large piece of butcher paper
- ▶ Crayons and markers
- ▶ Various pieces of clean trash
- ▶ Tape
- ▶ Book *Wartville Wizard* by Don Madden

Lesson 10

- ▶ A penny or dime per student

Lesson 11

- ▶ Tarp or blanket
- ▶ Several cups of dry cereal

Lesson 14

- ▶ *If You Hopped Like a Frog* by David Schwartz

Lesson 15

- ▶ Box of colored toothpicks
- ▶ *Chameleons are Cool* by Martin Jenkins
- ▶ The poem called *The Chameleon* from the book *Zoo Doings* by Jack Prelutsky
- ▶ Various colors of construction paper
- ▶ Scissors
- ▶ Glue
- ▶ Markers

Lesson 19

▶ Individualized list of materials for students to construct their zoo exhibits

Lesson 22

▶ One piece of posterboard
▶ Marker

Lesson 27

▶ Blank paper or posterboard for each small group of students

Lesson 28

▶ 3 × 5 card for each student

About the Author

Jason S. McIntosh, Ph.D., is an experienced educator (25 years in the field) and passionate advocate for gifted education. He earned his doctorate in Gifted, Creative, and Talented Studies at Purdue's Gifted Education Research Institute in 2015 and is currently serving as a gifted coordinator and independent consultant. Since his time at Purdue, Jason has authored seven NAGC Curriculum Studies Network Award-winning curriculum units (2016–2022) and hopes to write many more in the future. Other past projects include working with Dr. Marcia Gentry to identify high-potential Native American youth from the Diné, Lakota, and Ojibwe tribes, serving as the president of the Arizona Association for Gifted and Talented (AAGT), and earning the Feldhusen Doctoral Fellowship Award in Gifted Education. To find out more about his latest curriculum projects, please visit his website at www.notmoreofthesame.com.

Common Core State Standards Alignment

ELA	Reading Literature	RL.3.1 Ask and answer questions to demonstrate understanding of a text, referring explicitly to the text as the basis for the answers.
		RL.3.2 Recount stories, including fables, folk-tales, and myths from diverse cultures; determine the central message, lesson, or moral, and explain how it is conveyed through key details in the text.
		RL. 3.5 Refer to parts of stories, dramas, and poems when writing or speaking about a text, using terms such as chapter, scene, and stanza; describe how each successive part builds on earlier sections.

	Reading Informational Text	RI.3.1 Ask and answer questions to demonstrate understanding of a text, referring explicitly to the text as the basis for the answers.
		RI.3.3 Describe the relationship between a series of historical events, scientific ideas or concepts, or steps in technical procedures in a text, using language that pertains to time, sequence, and cause/effect.
		RI.3.4 Determine the meaning of general academic and domain-specific words and phrases in a text relevant to a Grade 3 topic or subject area.
		RI.3.9 Compare and contrast the most important points and key details presented in two texts on the same topic.
	Writing	W.3.2 Write informative/explanatory texts to examine a topic and convey ideas and information clearly.
		W.3.7 Conduct short research projects that build knowledge about a topic.
		W.3.8 Recall information from experiences or gather information from print and digital sources; take brief notes on sources and sort evidence into provided categories.
	Speaking and Listening	SL.3.1 Engage effectively in a range of collaborative discussions (one-on-one, in groups, and teacher-led) with diverse partners on Grade 3 topics and texts, building on others' ideas and expressing their own clearly.
		SL.3.4 Report on a topic or text, tell a story, or recount an experience with appropriate facts and relevant, descriptive details, speaking clearly at an understandable pace.
Math	Number and Operations – Base Ten	3.NBT.A Use place value understanding and properties of operations to perform multi-digit arithmetic.
	Measurement and Data	3.MD.B Represent and interpret data.
	Geometry	Geometry 3.G.A Reason with shapes and their attributes.

Next Generation Science Standards Alignment

Grade 2	2-LS4-1. Make observations of plants and animals to compare the diversity of life in different habitats.
	K-2-ETS1-2. Develop a simple sketch, drawing, or physical model to illustrate how the shape of an object helps it function as needed to solve a given problem.
Grade 3	3-LS1-1. Develop models to describe that organisms have unique and diverse life cycles but all have in common birth, growth, reproduction, and death.
	3-LS2-1. Construct an argument that some animals form groups that help members survive.
	3-LS3-2. Use evidence to support the explanation that traits can be influenced by the environment.
	3-LS4-3. Construct an argument with evidence that in a particular habitat some organisms can survive well, some less well, and some cannot survive at all.
	3-ESS2-2. Obtain and combine information to describe climates in different regions of the world.

Grade 4	4-LS1-1. Construct an argument that plants and animals have internal and external structures that function to support survival, growth, behavior, and reproduction.
	4-LS1-2. Use a model to describe that animals receive different types of information through their senses, process the information in their brain, and respond to the information in different ways.
Grade 5	5-ESS3-1. Obtain and combine information about ways individual communities use science ideas to protect the Earth's resources and environment.
Grades 3–5	3-5-ETS1-1. Define a simple design problem reflecting a need or a want that includes specified criteria for success and constraints on materials, time, or cost.
	3-5-ETS1-2. Generate and compare multiple possible solutions to a problem based on how well each is likely to meet the criteria and constraints of the problem.